Eco-disasters

Seok Soon Park

Eco-disasters

What Happened and How to Prevent Them

Springer

Seok Soon Park
Department of Environmental Science
and Engineering
Ewha Womans University
Seoul, Korea (Republic of)

ISBN 978-981-96-8666-7 ISBN 978-981-96-8667-4 (eBook)
https://doi.org/10.1007/978-981-96-8667-4

Translation from the Korean language edition: "Environmental Disasters and Human Survivorship" by Seok Soon Park, © Amoonhaksa 2020. Published by Amoonhaksa. All Rights Reserved.

The original submitted manuscript has been translated into English. The translation was done using artificial intelligence. A subsequent revision was performed by the author(s) to further refine the work and to ensure that the translation is appropriate concerning content and scientific correctness. It may, however, read stylistically different from a conventional translation.

This Springer imprint is published by the registered company Springer Nature Singapore Pte Ltd.
The registered company address is: 152 Beach Road, #21-01/04 Gateway East, Singapore 189721, Singapore

This book is dedicated to these three great professors

Dr. Christopher G. Uchrin of Rutgers University
and Dr. Peter R. Jaffé of Princeton University
who have been of great help to my academic journey
and
Dr. Michael R. Reich of Harvard University

who played a crucial role in shaping my hobby into this global edition.

Foreword

Today, when awareness and understanding of environmental and ecological issues are more important than ever, Prof. Seok Soon Park's book *Eco-disasters: What Happened and How to Prevent Them* pinpoints the seriousness of environmental disasters and the need to prepare for them in a timely manner. This book realistically describes various environmental knowledge and lessons that we not only need to know but also to master by examining cases of past environmental disasters and mistakes. It will be a valuable resource to prevent environmental disasters in advance, reduce damages and harms when they occur, and foster a better future environment for all.

Professor Park became deeply interested in environmental disasters when he started graduate studies in the United States in the early 1980s. At that time, the Three Mile Island nuclear power plant accident and the Love Canal's industrial dumping of toxic chemicals and soil pollution incident became a big social alarm-bell, and through these events, universities in Korea and elsewhere began to educate and study the seriousness and importance of environmental disasters. Prof. Park was interested in various lessons of environmental disasters by teaching them in his classes and wanted to find out more about prominent cases. His interest soon developed into a hobby of collecting and synthesizing all available data on environmental disasters.

Since then, he has collected and studied various common and unique kinds of environmental disasters around the world. In 1994, Prof. Park first published a small book called "Environmental Disasters in Global Villages" based on the Bhopal Gas Leak tragedy in India, the Chernobyl Nuclear Disaster in Ukraine, the Rhine River Basel Chemical Leak in Switzerland, etc., that he had collected until then, and has since continuously improved and supplemented its contents until today.

Environmental disasters have increased with the development of industrial civilization, and indeed the general public from the lowest economic ladder often suffers the greatest harms and damages. As a result, anti-industrial and anti-capitalist environmental movements and labor movements frequently cite instances of such disasters to score social and political points. However, Prof. Park did not write this book for that purpose. His book focuses on correctly understanding past disasters and preparing for a safe future following the best scientific principles and practices. In particular, he strives for an affluent and enriched environment based on human dignity that allows all mankind to enjoy healthy, pleasant, and prosperous lives. In addition, he promotes a scientific environment that prioritizes rational thinking and logical alternatives as a strategy to prepare for most preventable accidents and disasters.

This book systematically covers cases of environmental disasters by dividing them into seven parts. Each part details the background, development process, outcomes, and follow-up measures of various types of disasters around the world from the 19th century to the present. His approach will permit readers to have a scientific and comprehensive understanding of environmental disasters. In addition, the book provides important insights and essential guidelines on what to do in order to prevent environmental disasters. Therefore, the content of the book will be of great benefits to various readers, including governments, businesses, environmental experts, and the general public.

Lastly, I would like to thank the Springer Nature for introducing Prof. Park's long-standing research and writing activities to overseas readers beyond Korea alone. I hope that this book will be widely read abroad, and help protecting the environment, preventing disasters, and contributing greatly to the healthy life and prosperity of all mankind. I commend Prof. Park for his outstanding environmental research and dedication, and highly recommend this book.

Myung Bak Lee
17th President of the Republic of Korea
Office of the Former President,
Seocho-Gu, Seoul, Korea (Republic of)

Preface

Learning from the Past Can Prevent Future Disasters

I first became interested in environmental disasters in the early 1980s when I began my graduate studies in the United States. At that time, the Three Mile Island nuclear power plant accident and the Love Canal dumping of toxic chemicals had recently occurred, and their impacts were still being felt. Consequently, environmental disasters were frequently cited in lectures and became an important social issue. Through lectures I attended, I encountered various stories of environmental disasters, which sparked my desire to learn more about these cases in detail.

My interest began in this way had since evolved into a hobby of collecting and organizing materials related to environmental disasters. Believing that learning from the past could prevent future disasters, I started gathering stories and photos scattered throughout books and newspapers whenever I had the time. Unfortunately, new disasters kept recurring, which aided my hobby! Major incidents that shocked the world continued to happen: the Bhopal disaster in India in 1984, the Chernobyl disaster in Ukraine in 1986, and the Rhine River Basel spill in Switzerland in 1986. During the 1980s, an unusually high number of environmental disasters continued to manifest worldwide.

After returning to South Korea in 1988, I introduced the stories of environmental disasters that I have collected through newspapers and media broadcasts, and I wrote a small book called *Environmental Disasters in the Global Village* in 1994. From 1994 to 1995, I spent a sabbatical year at Princeton University in the United States, where I collected and organized even more episodes of environmental accidents and disasters. During that time, I discovered "Island of Dreams: Environmental Crisis in Japan" in Princeton's Firestone Library, and I couldn't contain my excitement, feeling like I had struck gold. By encountering this book, I was able not only to grasp the exact facts of various disasters in Japan, but my hobby expanded into a new area of translating world-class books into Korean. So far, I have published more than ten translated books, which has indeed made me a lifelong learner.

"Island of Dreams" detailed the modern environmental history of Japan, where numerous environmental problems had occurred arising from rapid, not-well-planned industrialization on a small landmass with a high population. Given that South Korea was following Japan's industrial footsteps, I believed there was much to learn from them. I contacted Prof. Michael Reich of Harvard University, the co-author of the book, and published the Korean translated version in 2001.

In 2003, our University Press proposed that I create an educational comic book. So I selected 21 major incidents and published this as a very engrossing comic book in order to bring awareness to wider audience in Korea. It was fortunate that the environmental disaster cases I had collected through my long-standing hobby and obsession could be shared with the public through various channels. Throughout these times, my hobby continued, and new environmental disasters recurred. In 2005, I published a revised version with more content, after adding several new disasters. The revised edition immediately became a bestseller, being selected as the "Youth Book of the Year" by the Korean Publishing Culture Association in December 2005 and as the "Recommended Book of the Month" by the Publication Ethics Committee in January 2006.

Around 2011, I left the university for a while and served as the president of the Korea National Institute of Environmental Research, since environmental disasters have been occurring more frequently in South Korea. I personally experienced the severity of these disasters on a different level as a relevant public official, with incidents such as the detection of radioactive asphalt on the roads in the Seoul metropolitan area in November 2011, the hydrofluoric acid leak at Huber Global in Gumi in September 2012, and the hydrochloric acid leak at Woongjin Polysilicon in Sangju in January 2013. In particular, the 2012 hydrofluoric acid leak was a major incident, resulting in the deaths of 5 workers, injuries to 18 firefighters, hospital treatment for approximately 12,000 residents, the deaths of about 4000 livestock, the destruction of 212 hectares of crops, and compensation payments to residents alone amounting to 28 million dollars. I reported the details of this incident directly to the president during a Cabinet meeting at the Blue House and answered questions. Additionally, I had to answer criticisms from Members of the National Assembly regarding the measures taken in response to the incident.

After returning to the university, I devised a new graduate course called "Case Studies of Environmental Disasters." This is because I thought that environmental disasters was a field that required systematic education and research beyond the stage of collecting data as a hobby and writing popular science books. Consequently, I have written this book to be used as a textbook for this newly designed graduate educational course. As an author, I was very satisfied to see that students began to think and appreciate about the environment through a new perspective and found it interesting and challenging. This reaction from students stimulated my enthusiasm for penning this book once again.

In 2020, the global COVID-19 pandemic led to the beginning of preventative measures and afforded me more time, I added new incidents and enriched the content, resulting in the publication of yet another revised edition. Moreover, I wanted to publish an English version and introduce it worldwide when I had the opportunity. I believed this book was unique in comprehensively covering various types of environmental disasters worldwide, explaining the background, progression, outcomes, and post-disaster measures. Furthermore, I thought that the refined content developed over time would be well-received even when introduced abroad. Finally, my latest dream of introducing the environmental disaster books I have authored over the past 40 years to international readers has come true thanks to the Springer Nature.

This book consists of seven parts covering 42 representative environmental disasters. Part I introduces representative disasters in Europe, America, and Asia in the nineteenth century. Part II covers disasters related to the atmosphere. Part III deals with disasters caused by water and food affected by heavy metals, toxic chemicals, and pathogenic microorganisms. Part IV examines poisoned environment caused by toxic pollutants and oil spills. Part V introduces nuclear power plant accidents and industrial chemical leaks. Part VI deals with disasters occurring in the modern living environment, such as indoor air, radon, and electromagnetic radiation, and Part VII presents cross-border disputes and planetary-wide concerns. Furthermore, this book introduces nearly all disasters that have occurred to date by adding and grouping them into similar cases alongside representative environmental disasters of each type. For example, disasters like the Exxon Valdez oil spill are detailed in parallel with similar disasters such as the Amoco Cadiz oil spill. Similarly, industrial air pollution disasters like the Meuse Valley disaster are described within the context of events like the Donora disaster.

Environmental disasters often begin with minor human mistakes or ignorance, but the scale of the damage can be enormous, which may sound shocking to the general public. Moreover, the victims are often involved large numbers of innocent civilians, while the perpetrators are sometimes reckless and unethical corporations. Due to these characteristics, environmental disasters are frequently used by anti-capitalist environmental and labor movements to support their various causes. However, it's important to stress that the purpose of this book is not geared towards such applications. The purpose of writing this book is to accurately understand the environmental disasters that have occurred in the past and to be prepared and minimized all avoidable problems for the future. Specifically, governments, corporations, and environmental experts, who bear direct responsibility for environmental disasters, should not regard these events as merely absurd or shocking incidents from the past. Instead, they should analyze them scientifically and learn from them. Additionally, rather than instilling fear in young people or the general public, I hope this book will highlight the importance of the environment, contributing to the creation of a more eco-friendly, bright and healthy society.

Many people have contributed to the creation of this book. First, I would like to thank the many readers in South Korea who have recognized the importance of this book and read it avidly. I am also grateful to Amoonhaksa, which published multiple revised editions domestically, and to the Ewha Womans University Press, which created the comic book version. In addition, I would like to express my gratitude to Mr. Smith Ahram Chae, senior editor of the Springer Nature, for proposing the English edition, Mr. David Craig, British author of "There Is No Climate Crisis," for advising on the book title and the closing remarks, and Dr. Willie Soon, the founder of CERES (Center for Environmental Research and Earth Sciences) in the United States for the final proof reading. Finally, I sincerely thank the 17th President of Republic of Korea, Mr. Myung Bak Lee, who wrote the foreword for this book in the hope that many good Korean books will be read abroad.

Seoul, Korea (Republic of) Seok Soon Park

Competing Interests The author has no competing interests to declare that are relevant to the content of this manuscript.

Contents

Part I
Before the Twentieth Century

Chapter 1
London Cholera Outbreak

London, UK was once the top maritime trading base in Europe. From the Middle Ages, ships loaded with goods produced inland would follow the Thames River to the sea, heading towards Scotland or Ireland. When they returned after selling the goods brought from London, they loaded coal produced there as ballast to balance the empty ship and unloaded it at the dock. What was initially just a useless black stone, coal was used as heating fuel and by the late eighteenth century, it became the fuel for steam engines, the catalyst for the Industrial Revolution.

With the start of the Industrial Revolution in the UK, many people flocked to London. By the early nineteenth century, the UK had developed its industry to the point where it accounted for one-third of the world's trade volume. By the time the twentieth century began, London was the world's largest city with a population of 6.6 million. However, rapid industrialization and urbanization led to serious environmental problems. The Thames River stank from industrial waste and domestic sewage, and the sky was filled with smog due to the coal burned in factories and homes.

The first environmental disaster in London after the Industrial Revolution was a cholera outbreak, which claimed many lives. In the 1853~1854 epidemic in London over 10,000 people died and there were 23,000 deaths for all of Great Britain.[1] Originally, cholera was a contagious disease that spread widely in Asia, including India, China, Korea, and Japan, in the early nineteenth century. After that, cholera spread to Europe around 1823~1824 through Russia. As London was the world's largest maritime trading base at the time, it was a place where many foreigners travelled making it easy for various infectious diseases to spread.

At that time, London was severely lacking in urban infrastructure due to its rapidly increasing population, and untreated domestic sewage was discharged, contaminating even the well water used for drinking. However, with the environmental knowledge and medical level that the British had at the time, they could not know the route of

[1] https://en.wikipedia.org/wiki/1846-1860_cholera_pandemic.

S. S. Park, *Eco-disasters*, https://doi.org/10.1007/978-981-96-8667-4_1

pollution or the cause of the disease, so no measures could be taken. When cholera first broke out, people who had never experienced symptoms of severe thirst, physical decline, loss of consciousness, and eventually death, believed that this disease was due to the dirty city environment of London. Cholera was thought to be caused by polluted air, as it first started in the city center where air pollution was severe.

In 1854, when another cholera outbreak occurred, the cause was revealed. That summer, cholera began to rage in the center of London, and in just ten days, over 500 people died. John Snow, a then physician and surgeon, had been observing London's ongoing cholera for the past decade with a unique perspective.[2] 2 He believed that cholera was not caused by dirty air as the general public believed, but was transmitted through contaminated drinking water. Snow thought that if air or fog were the source of contamination, the outbreak areas should be evenly distributed throughout the city. But in reality, patients were mainly concentrated around Broad Street, and he began to mark the occurrence of cases on a map.

He thought that if the air was the cause when cholera broke out in 1848, patients should feel pain in their lungs. But since they complained of pain in their intestines, he suspected that drinking water was the problem. While tracking the commonalities of patients, Snow confirmed that the water they drank came from a well on Broad Street, now called Broadwick Street. Also, he found out that the first place where cholera broke out was the home of a sailor who travelled overseas, and that there were many deaths in poor houses directly used water from the Thames River.

From these facts, Snow was able to confirm that cholera was transmitted through drinking water and urged the city of London to remove the pump handle from the well on Broad Street to prevent people from drinking that water. Initially, the relevant officials who did not believe his claim accepted it as the situation became urgent. Following this the number of cholera cases rapidly decreased. Through this, Snow left a legacy of establishing the foundation of modern epidemiology by systematically revealing the interrelationships between sewage discharge points, drinking water intake points, and outbreak areas. Also, unlike in the past when there were tens of thousands of victims during cholera outbreaks, in 1854, thanks to Snow's contribution, only about 2,000 people died and cholera was eradicated. After this incident, a water supply committee was formed in London and scientific water management was carried out. Standards for managing drinking water were established, and dirty ponds and toilets around water sources were removed. It was also prohibited to dump human and animal excrement and waste from slaughterhouses into the river.

After this incident, a water supply committee was formed in London and scientific water management was carried out. Standards for managing drinking water were established, and dirty ponds and toilets around the water source were removed, and it was also prohibited to dump human and animal excrement and waste from slaughterhouses into the river.

A few years later, in the steaming hot summer of 1858, the 'Great Stink' occurred in London.[3] That summer, human excrement released into the Thames River caused

[2] https://en.wikipedia.org/wiki/john_snow.

[3] https://en.wikipedia.org/wiki/great_stink.

a foul smell to spread across the capital. The boom in popularity of flushing toilets (marketed to the masses at the Great Exhibition in 1851) only made things worse, overwhelming old cesspools and forcing ever more effluent into the river. Many of the rich fled the city. Even the Houses of Parliament were closed. Subsequently, in 1865, sewer pipes were installed throughout London for the first time in a modern city, and in 1876, the world's first 'River Pollution Prevention Act' was enacted and promulgated. As part of the efforts to save the Thames at the time, a method of installing large sewage pipelines on both sides of the river and dumping sewage into the sea was also attempted. But due to the Thames's tides the effluent flowed back into the city at each high tide.

In 1878 two ships, the Princess Alice and the Bywell Castle collided on the Thames, causing many deaths. It was claimed that the high death toll was because the polluted river water contained a lot of hydrogen sulfide.[4] This ship collision sparked national interest in Thames pollution. In 1882, the Metropolitan Sewage Disposal Committee was formed to identify the cause of the stench from the Thames and seek sewage treatment methods. Around this time, a scientist named William Dibdin conducted the first systematic study of water pollution in the Thames, measuring ammonia and dissolved oxygen. This work introduced the concept of Biochemical Oxygen Demand (BOD) to explain the degree of water pollution, and raised the need for a new sewage management system.[5] Dibdin suggested improving the sewage management system in London at the time, purifying sewage through filtration and sedimentation, and installing sewage outlets closer to the sea.

In 1889, the world's first method of sedimentation treatment of organic matter included in domestic sewage (primary treatment) was developed in the UK. As a result, by around 1891, the water quality of the Thames had greatly improved and the fish that had disappeared started to return to the river. In 1893, the 'Trickling Filter Method', one of the biological treatment methods (secondary treatment) that decomposes organic matter using microorganisms, was developed and used for the first time in the world. The widely used biological treatment method 'Activated Sludge Method' was also first developed in the UK in 1914.

Such efforts made in the UK have greatly contributed to the development of environmental knowledge and technology today. The process of water purification in rivers, the importance and measurement methods of underwater oxygen, and the academic methods of expressing the degree of pollution due to organic matter. The basic theory was established around London and the Thames at this time. However, such efforts did not yield any significant results, being swept away by the global expansion, industrial expansion, and wars that Britain was pursuing at the time, and the Thames turned back into a cesspool.

After the end of World War II, the population of London increased to over 8 million, and more pollutants flowed into the Thames, reaching the worst water quality around 1950. In 1957, all the fish in the river disappeared. The British Parliament, located on the banks of the Thames, could not open the windows due to the stench

---

[4] https://en.wikipedia.org/wiki/sinking_of_ss_princess_alice.

[5] https://www.amazon.co.uk/Purification-Sewage-William-Joseph-Dibdin/dp/114377597.

from the river water. The parliament prioritized supplying London citizens with water that did not smell bad and struggled to find a solution.

In the 1960s, the British government resumed the Thames Revival Movement. In 1961, it banned the dumping of waste and garbage into the river and started regional river purification projects. In 1963, the Water Resources Act was enacted to establish a dedicated river management department in each region. Also, sewage treatment plants were built all over the UK to ensure thorough water pollution regulations. In 1970, an independent administrative agency, the Environment Agency, was established to handle pollution issues, and a Royal Commission was also formed to prevent water pollution. A water quality monitoring network was established to maintain good water quality throughout the country. Thanks to these efforts, the Thames began to revive. Around 1974, salmon, which had disappeared from the Thames in the early nineteenth century, returned.[6] The Thames has now revived as a river in which fish thrive as does birdlife like swans in addition to, providing citizens with safe drinking water and a pleasant space for relaxation.

The London cholera incident gave birth to epidemiology, which is the backbone of modern preventive medicine, and was an opportunity for cities in Europe at the time to introduce scientific sanitation methods. Also, the history of the Thames, which was abandoned as a river of death after the Industrial Revolution and revived, is cited as the best example proving the U-Turn theory of the environment. The U-Turn theory is that while pollution is aggravated and the quality of the environment deteriorates during the initial industrialization, as economic growth reaches a certain level, people's interest in the environment increases and technology improves, and the environment recovers. The U-Turn process of the Thames has left humanity with much knowledge of environmental science and technology.

---

[6] https://www.arabianbusiness.com/industries/industries-culture-society/salmon-do-U-turn-in-river-thames-until-sewers-fixed-11539.

Chapter 2
Chicago Typhoid Outbreak

Chicago is a city formed situated on the Great Plains and the Great Lakes in the Midwestern region of the United States. In the early nineteenth century, a military base called Fort Dearborn was built at the mouth of the Chicago River flowing into Lake Michigan, and the city that emerged as people settled here is Chicago. By the mid-nineteenth century, the population of Chicago had reached about 30,000, and all domestic sewage flowed along the Chicago River into Lake Michigan. Due to the small population at the time, this didn't cause any major problems.

After the end of the American Civil War in 1864, the city's population grew rapidly. Chicago, which experienced a major fire in 1871 which burned most of the city, grew even faster with the subsequent redevelopment project.[1] High-rise buildings were built in the city center, and the population increased rapidly, exceeding one million in 1885. As the population increased, the city expanded to the adjacent Calumet River basin, and the streets were overflowing with excrement and garbage. Moreover, increasing amounts of domestic sewage were discharged into the Chicago River and Calumet River, both of which flowed into Lake Michigan, polluting the water. The citizens of Chicago suffered from bad smells and diseases, and could no longer obtain clean drinking water.

In 1885, typhoid broke out, killing about 90,000 people.[2] It was a disaster caused by the water of Lake Michigan, where the sewage and garbage of one million people were dumped. This was the event with the most deaths among the environmental disasters recorded so far. Even after that, epidemics in Chicago continued due to the polluted rivers and lake. According to records, about 20,000 Chicago citizens contracted typhoid again from 1891 to 1895. As long as Chicago used the water from Lake Michigan and dumped sewage back into it, the city's inhabitants would continue to suffer from waterborne diseases.

[1] https://en.wikipedia.org/wiki/great_chicago_fire.

[2] https://en.wikipedia.org/wiki/Chicago_1885_cholera_epidemic_myth.

S. S. Park, *Eco-disasters*, https://doi.org/10.1007/978-981-96-8667-4_2

After repeated outbreaks of typhoid, the City of Chicago planned a large-scale civil engineering project to change the direction of the Chicago River and the Calumet River, which used to flow into Lake Michigan, to flow into the Illinois River, a tributary of the Mississippi River.[3] This reversed the flow of the two rivers polluted by Chicago's sewage. Thanks to this change, Chicago's sewage no longer flowed into Lake Michigan, but instead followed the Mississippi River through St. Louis to New Orleans downstream. With the completion of construction in the early twentieth century, the water quality of Lake Michigan improved significantly and the repeated epidemics ceased.

However, the largest civil engineering project since the founding of the nation, which diverted the flow of polluted rivers to drive out epidemics and save Lake Michigan, was embroiled in lawsuits with the two regions which were damaged as a result. First, several cities, including St. Louis located downstream of the Illinois and Mississippi Rivers, which received Chicago's sewage, sued the city of Chicago and the state of Illinois. Also, the states of Canada and the United States located on the coast of Lake Michigan sued Chicago for drawing a large amount of water from Lake Michigan and not discharging it back, causing the lake level to drop. In fact, after the change, the levels of Lake Michigan and Lake Ontario were found to have dropped by about 20~30 cm. Such backlash against the change was mitigated as the City of Chicago increased groundwater use and discharged wastewater after purification with sewage treatment technology that became available at the time.

The change in the direction of river flows first attempted in Chicago was also carried out in several other regions. Around 1930, when Seattle's domestic sewage flowed into adjacent Lake Washington causing serious water quality deterioration, the city of Seattle changed the flow of domestic sewage, that had been going into Lake Washington, to Puget Sound on the Pacific coast in 1936.[4] Around 1940, as the city expanded to the suburbs, the water quality of Lake Washington became a problem again around 1955. In 1963, the city of Seattle made domestic sewage from the suburban areas also flow into Puget Sound, improving the water quality of the lake. Lake Geneva (also known as Lac Léman) in Switzerland, which experienced serious lake eutrophication with the generalization of soap use in the late 1950s, also improved the water quality of the lake through the construction of sewage treatment plants and changes in the water catchment areas of the inflowing rivers.[5] However, as seen in the case of Chicago, changes in the direction of rivers can cause inter-regional disputes. Especially in times like now, when there is a heightened understanding of water shortage and water pollution between regions.

The Chicago disaster served as a driving force to advance America's water treatment and sewage treatment technology. At that time, there were frequent outbreaks of epidemics nationwide in the United States. The population was increasing and urbanization and industrialization were rapidly progressing. But the main cause was that water supply and sewage treatment could not keep up with these trends. By the

---

[3] https://interactive.wttw.com/chicago-river-tour/how-chicago-reversed-river-animated.

[4] https://www.historylink.org/File/11250.

[5] https://www.cipel.org/en/le-leman/presentation-du-leman/.

turn of the twentieth century, diarrhoeal diseases caused by contaminated drinking water were the third leading cause of death in the United States after pneumonia and tuberculosis.

In 1887, for the first time in the United States, a device that filtered water drawn from a river through sand was installed in Lawrence, Massachusetts. After that, the incidence of typhoid in this area significantly decreased. As a result, many cities across the United States introduced filtration devices. Around 1907, 33 cities in the United States used mechanical filtration devices, and 13 cities used sand filtration devices. A few years after the introduction of filtration devices, chlorination began to be used. Chlorination was first introduced in Jersey City, New Jersey, in 1908 by a physician named John L. Leal.[6] Despite facing tremendous opposition to his attempt to add chlorine, a poison, into drinking water, he did not back down and put it into action. As a result, the number of bacteria significantly decreased, and it was proven that chlorination is very effective in producing safe drinking water. Since then, chlorination has spread worldwide and is still widely used.

Another problem for American cities at the end of the nineteenth century was garbage dumped into rivers. The garbage in the rivers flowed to the sea, polluting the coast. In 1899, the U.S. Congress enacted the 'Rivers and Harbors Act (also known as Refuse Act)' prohibited dumping garbage into rivers and seas.[7] However, this law only prohibited the act of dumping garbage and did not regulate the discharge of sewage or other water pollutants. In 1907, the United States was the first country in the world to introduce the concept of water quality standards, laying the foundation for scientific water quality management. But strong federal regulations were not implemented until the end of World War II.

IIn 1948, the 'Federal Pollution Control Act', which regulates the discharge of pollutants, was finally implemented, and this law has been revised to become the 'Clean Water Act', the basic framework for water quality management today. With this, the U.S. government regulated the discharge of pollutants into public waters, especially industrial wastewater and domestic sewage. The Clean Water Act mandated wastewater treatment to minimize the pollutants being discharged. With the government's establishment of microbial water quality standards and chlorine disinfection in wastewater treatment and purification processes, waterborne diseases significantly decreased. Diseases like typhoid and cholera, which are waterborne, have since become rare occurrences. Through these processes, by the mid-twentieth century, people were able to drink cleaner water than ever before.

However, in the late twentieth century, drinking water faced a new challenge from toxic chemicals. After World War II, the use of synthetic organic chemicals surged and the contamination of water sources with these chemicals became a serious issue. The discovery of dead fish with grotesque tumors in rivers and lakes increased public concern about chemical pollution, which greatly influenced the enactment of laws. Furthermore, as the dangers of toxic chemicals were scientifically identified, fears about cancer increased. Moreover, improvements in chemical analysis technology

[6] https://www.amazon.com/Chlorine-Revolution-History-Water-Disinfection/dp/1583219137.

[7] https://en.wikipedia.org/wiki/refuse_act.

made it clear that it was difficult to manage drinking water with laws enacted in the 1940s.

In response to these historical changes, the Safe Drinking Water Act (SDWA) was enacted in 1974. This law was the most comprehensive law managing water sources and drinking water at the time. The Environmental Protection Agency (EPA), established in 1970, redefined the federal government's role through this law by managing the quality of rivers, lakes, and groundwater, improving environmental standards and regulatory systems, and providing financial support to state governments for watershed management. Part of the Safe Drinking Water Act was also to monitor whether these matters were being properly implemented. In addition, this law established national water quality standards for harmful substances and actively utilized all possible technologies to manage water quality.

The Chicago typhoid outbreak, which resulted in the largest number of deaths from a single environmental disaster at that time, served as a starting point for the realization that the government needs to manage the quality of rivers and lakes to ensure safe and clean drinking water supply for citizens. Also, this incident was an important catalyst for creating laws and systems to manage the quality of rivers and lakes and for developing effective wastewater treatment and purification technologies. The outbreak left a trail of canals connecting Lake Michigan and the Mississippi River, and as a result, Chicago grew into the third-largest city in the United States thanks in part to its inland waterways.

Chapter 3
Ashio Mine Disaster

In 1610, a massive copper vein was discovered in Ashio, a village in Tochigi Prefecture, 150 km north of Tokyo, Japan. A small-scale copper mine operated here for nearly two centuries until mining was halted in 1790. With the Meiji Restoration in 1868, the mine reopened and steadily increased its production. Twenty years later, the Ashio mine accounted for 44% of Japan's total copper production and emerged as one of the largest companies in Northeast Asia.

As the Ashio mine grew, so did the waste from mining and refining, all of which was dumped into the nearby Watarase River. This river supplied irrigation water to farmers downstream and was a good fishing ground for fishermen. The river occasionally flooded, creating fertile farmland where farmers lived and grew rice. However, as the Ashio mine continued to grow, it pushed out the surrounding forests and expanded its refining facilities. The mountains became increasingly barren and the emissions from the refining plant made the remaining vegetation disappear.[1]

From early 1878, signs of copper pollution began to appear in the river. The water was contaminated with copper, causing numerous fish to die off, and people who bathed in the river began to develop skin problems. As the years went by, the severity increased, with symptoms of copper poisoning appearing in all living organisms–plants, animals, and humans. A report investigating the Watarase River at the time described the situation as follows: "The farmland became barren and dried up. An epidemic swept through, killing numerous livestock, and people could not escape disease and death. The victims of this disaster, which could not be called a natural disaster, were left in miserable pain without clothes to wear or food to eat."

In 1890, residents downstream of the Watarase River, believing the Ashio mine to be the cause of the disaster, petitioned the government to halt mining, but were rejected. The residents then collected river water and nearby soil to scientifically prove that the damage to crops and the human body was due to the mine and requested analysis from the Ministry of Agriculture and Commerce. However, the Ministry

---

[1] https://en.wikipedia.org/wiki/ashio_copper_mine.

S. S. Park, *Eco-disasters*, https://doi.org/10.1007/978-981-96-8667-4_3

of Agriculture and Commerce repeatedly told them that they could not perform the analysis. Eventually, professors from the Tokyo Imperial University began an investigation and revealed that the cause of the copper poisoning that had spread to 28 villages at the time was the Ashio mine. The Ashio mine soon became a national issue.[2]

In December 1891, Shojo Tanaka, a member of the newly formed Japanese Parliament representing a district in Tochigi Prefecture, publicly questioned the Ministry of Agriculture and Commerce about the Ashio mine issue.[3] Asserting that the role of the administrative department and the new parliament was to guarantee and promote the welfare and livelihood of the people, Tanaka harshly criticized the government for failing to properly enforce mining-related laws. "If the mining industry is harmful to the public welfare, the Ministry of Agriculture and Commerce should not grant mining permits," he boldly argued, representing the will of the residents. However, before the Ministry of Agriculture and Commerce could officially respond to Tanaka's question, the parliament ended its session due to budget issues. The Ministry of Agriculture and Commerce responded during the recess, but only vaguely stated, "There was agricultural damage in the Watarase area, but we still don't know the cause." The government took no action and the debate was indefinitely postponed until the next session.

Re-elected, Tanaka once again attacked the government's passive attitude. The new Minister of Agriculture and Commerce avoided a direct answer to his inquiry, submitting only a written response stating, "Compensation for damages should be resolved at the company level." Moreover, he argued that the government could not order a halt to the company's operations. Eventually, local administrators in Tochigi Prefecture formed a mediation committee to resolve disputes between the affected farmers and the mine operators. The farmers agreed to a compensation of about 7,000 yen, and the company promised to operate a mine wastewater treatment facility and prevent further river pollution. The compensation was a paltry sum compared to the estimated damage of 11 million yen, and the treatment facility had little effect. Between 1891 and 1897, the area of polluted farmland increased from 2,000 hectares to 4,000 hectares, and the affected areas expanded to five prefectures, from Tochigi to Tokyo.

At this time, Japan was in no position to pay attention to the suffering of the farmers. In 1894, Japan entered a war with China (the Sino-Japanese War), and the demand for copper rapidly increased. If it was for Japan's victory, even the greatest sacrifice had to be endured. It was natural that productivity would decrease if investment was made in pollution control facilities, reducing the amount of copper produced for the war. The farmers also thought it was their duty not to bother the government, understanding the war situation. At that time, the copper production of the Ashio mine accounted for half of Japan's total production. The farmers

---

[2] Island of Dreams: Environmental Crisis in Japan, Norrie Huddle and Michael Reich, Schenkman Books, 1987.

[3] https://en.wikipedia.org/wiki/Shojo_Tanaka.

silently endured their land being polluted, vaguely hoping that compensation for their patriotic sacrifice would be awarded after the war.

In the fall of 1896, two major floods occurred, covering the surrounding farmland with severely contaminated river sediment. In February of the following year, when the war with China that had lasted for two years ended in victory, Representative Tanaka again demanded the government halt mining operations. However, the government claimed that the issue had already been resolved. Angered by the government's attitude, over 800 farmers stormed the Ministry of Agriculture and Commerce in Tokyo to protest. The government formed the Ashio Mine Pollution Investigation Committee and promised to visit the Ashio mine instead of the Ministry of Agriculture and Commerce to investigate the problem directly. But the situation did not improve at all.

One month after the first protest, in March 1897, over 3,000 farmers gathered in Tokyo for the second protest. The government responded by promising to prevent further floods caused by polluted water and to reduce taxes on the damaged farmland. As a result, a flood prevention embankment was built along the river, which seemed effective for a while. But the result was disappointing. The great flood of 1899 inundated areas that had not been affected before.

In February 1900, over 3,000 victims tried to protest again in Tokyo. However, they were stopped by armed police sent by the central government before they could even cross the current border. And a one-sided fight ensued. "Five policemen rushed in and lifted an old man up and threw him into the river. The faces of the protesters were covered in mud and their mouths were clogged with sand and filth. Injured people lay dead on both sides of the road, and the road was stained with a long trail of blood over 10 km." Sixty protest leaders were arrested and the trial lasted for two and a half years. The case was eventually dismissed, but during that time, the mining operation did not receive any legal sanctions.

In February 1900, over 3,000 victims tried to protest again in Tokyo. However, they were stopped by armed police sent by the central government before they could even cross the prefectural border. One-sided fight ensued. At the time, the newspaper reported as follows: "Five policemen rushed in and lifted an old man up and threw him into the river. The faces of the protesters were covered in mud and their mouths were clogged with sand and filth. Injured people lay dead on both sides of the road, and the road was stained with a long trail of blood over 10 km." Sixty protest leaders were arrested and their trial lasted for two and a half years. The case was eventually dismissed, but during that time, the mining operation did not receive any legal sanctions.

At that time, the Ashio mine disaster became one of the most urgent social issues in Japan. People gathered to form a pollution investigation committee and the Women's Christian Association formed a relief organization. Christian social activists held speeches and meetings day after day and, despite the Ministry of Education's prohibition, even students joined the protest. During the protest period, Representative Tanaka, who had converted to Christianity, spearheaded the Ashio movement. He was disillusioned with the evasive attitude of the parliament and the government and finally resigned from all public offices in October 1901 when he turned 61. Two months later, Tanaka, who joined the struggle as a citizen, tried to protest directly to the royal family instead of fighting the parliament.

In March 1902, there was a movement to establish a Mine Pollution Committee in the cabinet at the government level. However, this time a natural disaster that hindered this occurred. A large flood occurred in the Watarase River and clean soil from above the mine was washed down, covering the contaminated farmland, allowing farmers to cultivate crops again. The government and the mining company announced that the copper pollution problem had been resolved. As the superficial problem disappeared, the anger of the residents also subsided.

The government took advantage of the lull to announce plans to turn Yanaka Village, the center of the Ashio movement, into a reservoir for flood control. Tanaka moved to Yanaka Village and desperately resisted with the villagers. However, in the end, the 450 households living here sold their land to the government and migrated to Hokkaido, known as Siberia of Japan, in 1907. After that, the migrants lived with the belief that they would return to the land they had settled for hundreds of years. But as time passed, their hometown became a place they could not return to. The central government and the power of corporations had won. In the end, the Ashio Mine was operated until February 1973 and then stopped operations. Currently, the Ashio Mine has been restored and transformed into a tourist spot that vividly tells the story of that time.

This was the first recorded environmental disaster in Japan and the prelude to many similar incidents which followed. Minamata, Itai-itai, Niigata, Yokkaichi, Kanemi oil, and other disasters followed one after another on the Japanese archipelago, countless innocent people fell ill and died, and the entire country had to suffer due to development and pollution. The twentieth century of Japan in human history has forever remained a disgrace as a showcase of global environmental disasters.

Part II
Polluting the Air We Breathe

Chapter 4
Donora Smog

Donora, a small industrial town built on the metal processing and steel industry, is situated on the banks of the Monongahela River in Pennsylvania, USA. The town was established when William Donner started building a steel mill on the fast-flowing river in 1900, and people began to settle there. As the steel mill was completed, many immigrants who had migrated from Europe to the United States flocked to the area in search of jobs. Donora was a young town full of workers. It was a town where there were hardly any elderly people, and no deaths occurred for several months.

The steel industry was so closely intertwined with the lives of the local residents that even young children knew how steel was made. To make steel, iron ore, coal, and limestone are needed. At that time, one factory in this area used an average of 45 truckloads of iron ore, 40 truckloads of coal, 6 truckloads of limestone, and 6 truckloads of various raw materials per day. The amount of coal used here in a day was equivalent to the amount used in all households in the neighboring metropolis of Pittsburgh.

Coal was more important than anything else for the people of Donora. People used coal to keep their homes warm and to light the huge furnaces in the factories. The coal piled up like a mountain at the factories was a sign that the city's economy was running well. Residents overlooked the fact that their clothes were always covered in black dust, curtains and windows were dirty, and sometimes the sun was not visible all day. No one knew or cared about pollution. To them, coal dust was gold dust, and there was nothing more important than making a living.

In 1915, a zinc factory was built here. At the time, it was the largest zinc factory in the world. However, the newly built factory's chimney was only 45 m high, so it could not emit smoke above the surrounding mountains, which were about 180 m high. Despite the factory smoke settling on the ground, residents were satisfied with the fact that they received high wages even if they only worked three hours a day. The reason they only worked three hours a day was that no one could endure any longer. Although there were incidents of healthy young men fainting or quitting while working, people did not take these seriously.

S. S. Park, *Eco-disasters*, https://doi.org/10.1007/978-981-96-8667-4_4

Zinc is an essential element when present in small amounts in our bodies. But in large amounts, it is a substance that is highly toxic. A small amount of zinc combined with sugar helps fight the common cold. However, zinc combined with gases such as sulfur, carbon, fluorine, and nitrogen turns into a very dangerous substance. Generally, minerals containing fluorine and calcium are widely used in steel and iron smelting plants. During the refining process, these minerals produce highly corrosive fluorine gas, which can corrode glass and damage children's teeth.

The fluorine vapor emitted from steel mills, coal furnaces, and zinc blast furnaces was often trapped in the valleys due to the mountains surrounding the town of Donora. The surrounding mountains were devastated due to the smoke and heat emitted from the steel mills and zinc factories along the river. The steel mills and zinc factories flourished during World War II and the city was vibrant, but the shadow of death loomed ever darker.

On Tuesday October 26, 1948, a murderous fog took many lives in Donora.[1] That morning, Donora was covered in cold air. The air was cold and cloudy, and unusually windless weather caused all sorts of smoke from Donora's steel mills and furnaces to mix with the fog and fill the houses and streets. Even at noon, the darkness was so thick that it was almost impossible to distinguish anything, paralyzing all means of transportation. The air was yellowish and there was a disgusting smell.

The fog that started on Tuesday morning continued for several days and was motionless, blocked by the surrounding mountains. The residents thought this fog was just a weather phenomenon and focused on their daily lives. Three days later, on Friday afternoon, the annual Halloween parade in the United States was held as scheduled in the town. The ghost-like fog on the street created a perfect atmosphere for Halloween costumes. However, almost half of the town's residents were already sick and nine had died within a few days. Without knowing this, a high school football game was held on Saturday and the residents gathered to enjoy the game. Meanwhile, people continued to fall ill or die. By Sunday, when the fog began to clear, there were no coffins left at the funeral director's offices. On November 1, when rain fell and the fog disappeared, many of the town's citizens were busy holding funerals.

Twenty residents died in a week. After that, an additional 50 people died within a month, and as many as about 6,000 people fell ill. And its impact lasted long. Even after a decade, the mortality rate here remained significantly higher than in surrounding areas. After the incident, the health authorities temporarily halted operations at the zinc factory, but soon allowed them to resume, dismissing the smog as a one-time abnormal weather event. However, the smog in Donora was not a simple weather anomaly. The fact that most of the deaths occurred within 800 m of the zinc factory, where the smoke from the factory lingered the most, supported this.

Similar incidents had already occurred in Europe in 1930. From December 1 to 4, 1930, hundreds of patients had respiratory problems in the Meuse region of Belgium, and 63 people soon died from acute pneumonia and heart disease.[2] At that time, Belgium, like Donora, had established a large industrial zone focused on

[1] https://en.wikipedia.org/wiki/1948_Donora_smog.

[2] https://en.wikipedia.org/wiki/1930_Meuse_Valley_fog.

iron and steel industries in the valley region and continued operations, eventually suffering a serious environmental disaster. The valley of Meuse turned into a valley of death when the smoke emitted from the factories stagnated in the valley due to the calm and a phenomenon called 'a temperature inversion' which occurred in early winter. In meteorology, an inversion is a layer of warmer air held above colder air. Normally, air temperature gradually decreases as altitude increases, but in an inversion this relationship is reversed. An inversion traps air pollution, such as smog, close to the ground. At that time, not only people but also many livestock, birds, and trees around the valley were affected.

The Meuse Valley incident was widely known to most countries pursuing industrialization at the time, as it was the first major environmental disaster of the twentieth century. However, it seems that no one in Donora knew about the Meuse Valley incident. If only a little attention had been paid to the fact that Donora, like the industrial zone of Meuse, was in a valley and had long suffered from the smoke emitted from factories, much of the suffering could have been prevented. The disaster in Donora was like waiting for the Meuse Valley incident to be repeated.

The Donora incident played a decisive role in alerting the entire United States of the horrors of air pollution and in the federal government enacting regulatory legislation. The issue of air pollution in the United States had been a problem since the late nineteenth century. At that time, coal replaced wood and became the main energy source driving national economic growth. Locomotives and factory chimneys spewing coal smoke were symbols of modernization. In 1881, Chicago and Cincinnati enacted the first air pollution regulations in the United States to regulate the smoke and soot from furnaces and locomotives. By the time the twentieth century began, some local governments at the county level had also enacted pollution regulations. However, it took half a century for the state government to seriously consider pollution regulation. In 1952, Oregon was the first state to regulate air pollution by law. Other states soon followed suit.[3] The law enacted at this time primarily aimed to regulate smoke and dust.

In 1963, the United States enacted the 'Clean Air Act' at the federal level.[4] This Clean Air Act was a relatively weak law, supporting air pollution research financially and helping with cooperative programs within or between state governments. The year that marked a historical milestone in air pollution regulation in the United States was actually 1970. At this time, the Clean Air Act was revised into a strong law and the U.S. Environmental Protection Agency (EPA) was established, gaining the authority to set air quality standards for outdoor air nationwide and emission standards for air pollution applicable to cars. The Clean Air Act was revised again in 1990, further strengthening the authority of the Environmental Protection Agency and designating many pollutants of concern for toxicity as regulatory targets. It allowed the Environmental Protection Agency to regulate various low-concentration air pollutants classified as harmful to the human body. The list of low-concentration

[3] https://www.encyclopedia.com/environment/encyclopedias-almanacs-transcripts-and-maps/clean-air-act-1963-1970-1990.

[4] https://en.wikipedia.org/wiki/Clean_Air_Act_United_States.

toxic air pollutants included 188 substances, including benzene generated from gasoline, perchloroethylene emitted from dry cleaning facilities, and methylene chloride used as a solvent in some industries.

The Donora incident was a significant catalyst for the United States to achieve groundbreaking improvements in air quality. Emissions of pollutants across the United States have significantly decreased due to several revisions of the Clean Air Act and remarkable technological advances in the private sector. From 1970 to 1999, over approximately 30 years, the U.S. population increased by 33%, vehicle mileage by 140%, GDP by 147%, and coal usage for power production by about three times. But major air pollutants decreased by 31% during the same period. Sulfur dioxide, which played a fatal role in Donora, recorded its highest total emissions in the United States in 1972 at 32 million tons, but it steadily decreased afterwards, staying at 19 million tons in 1999, which was close to the emission level in 1915.

In 1998, the 50th anniversary of the Donora disaster, a monument was erected thanks to the efforts of a passionate high school student named Justin Shawley. The Pennsylvania Historical and Museum Commission erected a copper plate near the former steel mill to honor the victims.

Chapter 5
London Great Smog

On December 4, 1952, the weather in London, UK, suddenly began to get cold.[1] As is often the case with early winter weather in London, a high-pressure air mass formed in the southern region of the UK, causing a continuous calm with no wind, the sky was covered with clouds and a heavy fog covered the ground. Sunlight was blocked by clouds and fog, making it so dark during the day that it was hard to distinguish what was ahead, and the humidity exceeded 80%.

At the same time, a temperature inversion phenomenon occurred, where warm air settled above the cold air near the ground, preventing vertical mixing of the atmosphere and exacerbating pollution. Generally, the temperature decreases as you go up and smoke emitted by homes and factories, because it is hotter and thus lighter than the surrounding air, rises and gradually dilutes. However, when a temperature inversion occurs, where the atmospheric temperature near the ground is lower than the upper layer, the emitted smoke cannot rise and stays near the ground, causing damage to people's health.

As the temperature dropped to −1°C even at noon, fuel use rapidly increased throughout London. At that time, both homes and industries in the UK were using coal produced domestically as fuel. The smoke from burning coal was emitted into the atmosphere without filtration, and it stagnated on the ground without mixing with the upper atmosphere due to calm and temperature inversion phenomena. The emitted smoke combined with the thick fog to form smog, and in particular, the sulfur dioxide in the smoke turned into sulfuric acid. It was this sulfuric acid that had a fatal effect on the respiratory systems of London citizens.

This continued for a week until December 10. December 7 was recorded as the worst day, when thick smog covered almost all of London. During this period, the recorded soot concentration was about 16 times the average concentration in December, and the sulfur dioxide concentration increased more than 9 times. These figures quickly returned to normal as the wind began to blow on December 9. But

[1] https://en.wikipedia.org/wiki/great_smog_of_london.

S. S. Park, *Eco-disasters*, https://doi.org/10.1007/978-981-96-8667-4_5

during this period, a huge amount of pollutants, including about 1,000 tons of soot, over 2,000 tons of carbon dioxide, and about 370 tons of sulfur dioxide, were emitted into the thick fog every day.

However London citizens, who were always accustomed to thick fog, seemed only interested in the cancellation of various sports games, film screenings, and opera performances, even though the air was so dark that they could not distinguish what was ahead and traffic was on the verge of paralysis. Many newspapers published articles about smog. But they only covered everyday stories or economic losses. Not only ordinary citizens, but also doctors and government bureaucrats had no idea of the health consequences of what was happening, and there were no reports on diseases or deaths for a long time. Rather, the first people to recognize the harm that smog was doing to the human body were coroners, pathologists, and civil servants in charge of death reports. This was because the number of deaths suddenly increased dramatically during this period. The death rate at the time was much higher than any similar smog occurrence reported in the past.

As a result of severe air pollution for a week, 4,000 London citizens died from respiratory disorders and suffocation in the first three weeks after the incident. After that, an additional 8,000 people died from chronic lung diseases, resulting in a total of over 12,000 lives lost. The causes of death were mostly diseases of the respiratory, heart, and vascular systems. The deceased were mainly the elderly, children, patients, and other people with weak constitutions. There was also a surge in cardiopulmonary diseases in all age groups, with especially severe symptoms in those over 45. During the week of the incident, the citizens had no other measures to protect themselves than to leave London and take refuge elsewhere. Those who stayed had no choice but to wait for the atmosphere to recover through weather changes.

Historically London has had many reported cases of air pollution harm due to fog and coal. As early as 1273, Edward I was the first ruler in the world to issue a decree to reduce coal usage to prevent air pollution.[2] A similar decree was issued in 1306. It stated that violators would be sentenced to death though most people ignored the decree as coal was the only fuel for heating their homes. Only one person was ever executed. As we entered the industrial revolution, environmental disasters due to air pollution became more frequent. In 1878, 1880, 1891, 1892, 1911, 1948, 1952, 1956, 1957, and 1962, smog incidents occurred in London. Thus, London is recorded as the place where environmental disasters due to the same cause occurred most frequently in the same area.

Despite such frequent disasters, the British government was more concerned with avoiding the problem than preparing countermeasures. The quality of the air seemed to improve somewhat at the beginning of the twentieth century, but the smoke generated by coal combustion in homes was still a big problem. Therefore, whenever factories were accused of being the main culprits of air pollution, they argued that the smoke from the chimneys of homes in the capital accounted for 95% of the total air pollution emissions.

---

[2] https://www.thecmmgroup.com/history-air-pollution-part-1/.

The British government continued to evade responsibility and oppose new legislation until the disastrous event of 1952. At that time, the responsibility for air pollution was unclear among the relevant agencies and they tried to shift the blame to each other. Above all, the government was reluctant to bear the additional cost of supplying clean fuel in the difficult economic situation after World War II.

Even after experiencing tremendous damage in 1952, the British government was slow to take action. At that time, Britain was on the verge of a general election, so the government was just watching the situation, fearing that public opinion on their measures could pose a political threat. Government bureaucrats continued to do nothing, casually brushing off the ongoing deaths after the incident as due to an epidemic of influenza. As a desperate measure to avoid criticism for their lack of action, they recommended wearing masks, which in reality had little effect. Finally, in May 1953, they established the 'Beaver Committee', chaired by Sir Hugh Beaver, to study this issue and began to prepare measures. There were two main intentions behind this change in the government's attitude. First, 1952 was a year with many miners' strikes, and the conservative forces felt the need to create a new alternative to the situation of dependence on coal in energy policy, as the power of the miners' union was growing to the point of pressuring the government. Also, rumors began to circulate within the government in 1953 that another smog incident might occur. As pressure on the government continued from various quarters, the British government finally took action to solve the problem.

Based on the report submitted by the Beaver Committee, Britain enacted the 'British Clean Air Act' in 1956. This law was the first to deal with smoke, ash, and dust produced by domestic heating appliances, with the aim of preventing a recurrence of the smog incident. It also included support for the use of natural gas, oil, smokeless coal, and electricity for heating in homes instead of coal. But what was revolutionary was the introduction of a system to designate emission control areas, commonly known as 'Smokeless zones'. This system allows only approved fuels to be used in designated areas. This law greatly contributed to Britain's reduction of coal use and improvement of air quality.

In 1962, another serious smog incident occurred in London. This time, the weather conditions were very similar to the 1952 incident. But the smoke levels of the smog were about one third of those in 1952, although levels of sulfur dioxide were about the same. With the enactment of the Clean Air Act, the use of oil increased after 1957, and the use of natural gas also increased from the 1960s, while the proportion of heavy chemical industries decreased and power plants were relocated to the outskirts of the city. Only around 700 people died in the 1962 smog. The greatest force in this process was the relentless demand of citizens for cleaner air. The citizens' demand for clean air continued afterwards, and London currently maintains relatively clean air compared to most major cities worldwide.

The London smogs served as a wake-up call about the dangers of air pollution not only in the UK but also in countries around the world. Cities with a lot of fog are particularly sensitive to air pollution. So the use of energy sources which emit a lot of pollutants like coal should be suppressed. Once dangerous air pollution occurs, people can only wait helplessly until the polluted air is cleared by changes in weather patterns. Also, when thinking of air pollution, we usually first think of emissions from factory chimneys or cars. But the smogs show that heating fuel used in homes can also kill thousands of people.

Chapter 6
Los Angeles Smog

Los Angeles is the second-largest city in the United States. It is located in the desert climate zone of southern California. Generally, it is difficult for a metropolis to form in a desert climate due to the lack of water. But Los Angeles grew into the largest city in the western United States by securing water resources through the construction of artificial canals from the Owens Valley in the distant Rocky Mountains and the Colorado and Sacramento Rivers.

One of the reasons why a metropolis was built in the desert, where water is scarce in the United States, which has abundant natural resources and a vast land area, was due to the film industry that began to prosper from the early 1900s. At that time, lighting and filming technology were not developed, so intense sunlight was an important condition for film production. Los Angeles was able to produce films with clearer screens than any other region in the United States. So the American film industry flocked there and the city rapidly expanded. The American car culture, which started around this time, along with this intense sunlight, created a unique air pollution phenomenon in this area called Los Angeles smog.

The first report of symptoms of citizens suffering from respiratory difficulties and tears and eye pain due to the occurrence of brown smog in the Los Angeles atmosphere was in the summer of 1943.[1] The city of Los Angeles implemented various measures to solve this smog, but it did not achieve any significant effect. For several years, this brown smog worsened without the cause being identified, causing considerable damage not only to people but also to plants and livestock.

Professor Arie Haagen-Smit of the California Institute of Technology, who had been closely observing this phenomenon while living in Los Angeles, identified in 1951 that cars and intense sunlight were the cause of the brown smog.[2] He proposed the Photochemical Smog theory, which states that nitrogen oxides and volatile hydrocarbons emitted by cars generate toxic brown substances when exposed to intense

[1] https://laist.com/brief/news/climate-environment/the-heavy-history-of-smog-in-los-angeles.

[2] https://www.caltech.edu/about/news/fifty-years-clearing-skies-39248.

S. S. Park, *Eco-disasters*, https://doi.org/10.1007/978-981-96-8667-4_6

sunlight, and proved this through experiments. This smog was then distinguished from the previously known smog caused by sulfur dioxide and fog (London-type smog) and was called photochemical smog (Los Angeles-type smog).

While London-type smog forms as a dark gray on nights with weak sunlight and heavy fog, Los Angeles-type smog appears as a brown color during the day when the sunlight is intense. Because it is accompanied by substances like deadly ozone and Peroxyacetyl Nitrate (PAN), one should refrain from going out when the level is severe. Ozone, which does not dissolve well in water, penetrates deep into the lungs and causes inflammation and pulmonary edema even at low concentrations. In severe cases, it can lead to suffocation due to respiratory distress. PAN is a brown ultrafine dust that buries the city in the heat of the sun, causing reduced visibility.

Los Angeles, adjacent to the Pacific, exacerbates air pollution due to its geographical features. A high pressure that is permanently located in the eastern North Pacific covers the sky of Los Angeles, and temperature inversion often occurs due to the influence of the low-temperature sea breeze blowing from the coast. Therefore here, the inversion layer is always formed in summer and autumn. The pollutants emitted from at ground level cannot diffuse to the upper atmosphere and accumulate near the ground. Los Angeles has optimal weather and geographical conditions for the formation of photochemical smog, such as high pressure that presses the upper atmosphere to the ground, temperature inversion caused by warm winds blowing from the high-altitude desert, intense sunlight, and many calm windless days.

Los Angeles was the first city in the world to implement strong car emission regulations due to this unique photochemical smog. But the city still experienced serious air pollution due to increasing numbers of cars and the intense sunlight. In 1970, there were 148 ozone warnings in effect throughout the year. This warning is issued when the atmosphere is similar to the color of muddy water flowing from a river and, at this time, the elderly, children, and patients should stay indoors. After that, methods to reduce nitrogen oxides and hydrocarbons emitted from cars were studied, and in the 1970s, a new catalyst device was developed and commercialized, achieving some success in reducing photochemical smog. However, statistically the citizens of Los Angeles had a much higher rate of lung cancer than residents of other regions. In 1980, another serious photochemical smog occurred. During the first and second weeks of October that year, citizens were banned from jogging due to smog, and warnings were issued for children and the elderly to refrain from going out. During this period, many citizens suffered due to eye, ear, and respiratory diseases.

The sea breeze and temperature inversion due to high pressure are phenomena commonly seen in cities on the Pacific coast of the United States. California is currently the most affected state in the United States. Strong air pollution regulations are being implemented. The state government prepared an ambitious 20-year long-term plan in 1987 to eradicate the smog that has continued for decades. This plan was established under two basic concepts of legal regulation and citizen participation. In 1988, California enacted the Clean Air Act, which tightened the standards for automobile exhaust gas inspections and pollution industry regulations.[3] Companies employing more than 100 people were required to establish carpools, and a system was introduced to give parking permits preferentially to citizens who implemented this. To reduce vehicle operation, they are actively encouraging citizens' voluntary participation in working from home, reporting smoke-emitting vehicles, saving electricity, planting trees, and walking.

This policy was very successful and in 1997, ten years later, the ozone warning was issued only once. The fact that the ozone warning, which was issued almost every other day in 1970, was issued only once a year was a tremendous achievement. This was possible because of the development of environmental science and technology, industrial technology innovation, and the participation and cooperation of citizens. The United States has invested at least one trillion dollars for clean and clear air over the past 30 years. As a result, not only Los Angeles but also the air in New York, Chicago, Donora, and Pittsburgh, which were notorious for air pollution in the past, has cleared up again.[4]

The car culture that started in the United States in the last century brought about tremendous changes in human life. It became possible to easily supply goods and services far away, and gained the freedom to go anywhere at any time. Especially after World War II, as private cars became popular in developed countries, city dwellers were able to have homes in the suburbs with clean air and nature and enjoy vacation trips to their heart's content. Due to the freedom of movement that everyone desires, the number of cars has exploded over the past 100 years. The number of cars, which was only about 8,000 across the United States in 1900, increased to 170 million in a century.

Cars, which can be said to be one of the great achievements created by mankind, have caused many environmental problems around the world. Excessive energy consumption depletes limited oil resources, wide parking spaces and roads destroy nature, and toxic exhaust gases and traffic accidents threaten our health and lives. However, on the other hand, it is also true that thousands of tons of animal feces, which used to stink and teem with germs due to horse-drawn carriages in the past, have disappeared from the streets due to the advent of cars.

Mankind has been making relentless efforts to solve the environmental problems caused by cars, and the results have been very successful. The three-way catalytic converter (which converts CO, HC, NO to CO_2, H_2O, N_2) that emerged in the 1970s and the use of unleaded gasoline greatly contributed to reducing air pollution. Thanks to technological advancements, the amount of air pollutants emitted by a single car today has decreased to 1% compared to cars in the early 1970s. Recently, hybrid and electric cars have been commercialized, and hydrogen fuel cell cars have also started to appear on the roads. Therefore, environmental experts are very optimistic about

[3] https://digitalcommons.law.ggu.edu/cgi/viewcontent.cgi?referer=&httpsredir=1&article=1271&context=caldocs_assembly.

[4] https://www.epa.gov/clean-air-act-overview/progress-cleaning-air-and-improving-peoples-health.

the future air environment of advanced metropolises. However, the dust and noise generated on the road when when cars are running is still an unresolved problem.

Chapter 7
Yokkaichi Asthma

Yokkaichi on Honshu Island in Japan, located midway between Tokyo and Kyoto, has been industrially developed from an early stage due to its favorable port conditions. During the Meiji era, commerce and cotton weaving industries flourished, and during the early twentieth century invasion of East Asia, it was a frontline base for the military industry. In 1937, an oil refinery and a petroleum storage base for the Japanese navy were established here, and as a result, the entire city was reduced to ruins by U.S. bombing at the end of World War II.

After the war, Japanese industrial policy makers decided that the devastated Yokkaichi was very suitable for the petrochemical industry and began to establish industrial complexes. The local residents were very proud that their hometown was leading the new construction of Japan and did not spare their cooperation in the construction of the industrial complex. Starting with an oil refinery in 1956, various petrochemical plants moved in, and the industrial complex began operating in earnest from 1959.

As the industrial complex began to operate, local children started to catch colds more easily and had difficulty recovering. In the past, if they caught a cold, they would naturally recover by resting, but this new cold showed no signs of improvement. Residents began to cough one by one, and some had difficulty breathing. As the operating hours and intensity of the industrial complex increased, residents could not sleep due to the terrible stench and noise. Every time the wind blew from the direction of the industrial complex, the pungent smell of toxic chemicals stung their noses, and various respiratory diseases such as coughs, asthma, and chronic bronchitis tormented the residents. The city of Yokkaichi did not take any measures except to distribute only masks to the local residents.

Protests by residents began in April 1960, less than a year after the industrial complex began operating in earnest. As the residents' protests intensified, the city authorities established a Pollution Prevention Measures Committee in August 1960 and held its first meeting in October. The committee began investigating the state of air pollution and the health of residents. The number of asthma patients increased

S. S. Park, *Eco-disasters*, https://doi.org/10.1007/978-981-96-8667-4_7

dramatically in just a few months from spring to summer in 1961. According to a survey at the time, 48% of children under 10, 30% of the elderly over 60, and 19% of people in their 20 s showed respiratory abnormalities. The causative substances were identified as sulfur dioxide, nitrogen dioxide, and formaldehyde, but the air continued to be polluted indiscriminately because there was no way to stop the operation of the industrial complex and no technology to treat the pollutants. There was no way to avoid coughing unless you left Yokkaichi.[1]

The residents strongly demanded administrative measures to prevent pollution from the city authorities. The city authorities promised to provide medical services to asthma patients, asking them to bear with the inconvenience. Thus, they were able to use the hospital for free, but air pollution became increasingly severe and the residents had to suffer from diseases. The petrochemical complex using Middle Eastern crude oil containing a large amount of sulfur was emitting combustion gas from a chimney only 30 ~ 50 m high. Yokkaichi turned into a giant laboratory testing human limits.

Despite the situation, the city authorities continued to expand the scale of production and started operating the second petrochemical complex in 1963. The painful cough did not get better with any medicine, and the number of patients continued to increase. Free examinations and medical fee exemptions have been implemented since August 1962, but no compensation was made. At that time in Japan, there was no institutional mechanism to mediate disputes caused by environmental damage, so various social problems followed.

On April 2, 1964, on the third day of continuous smog covering Yokkaichi, 62-year-old Furukawa Yoshiro passed away. The cause of death was revealed to be lung cancer and chronic bronchitis. He was the first pollution disease death in Yokkaichi. Furukawa's death attracted national attention, and as a result, the city authorities built the world's first pollution disease management system at the end of 1964, not only in Japan. Even in the midst of this, the city authorities continued to establish the third complex and attract tenant companies.

On April 2, 1965, at the first anniversary memorial service for Furukawa, asthma victims formed the Yokkaichi Pollution Disease Patient Protection Association. They realized that they could not live quietly and endure as they had in the past. Receiving free medicine and being recognized as a pollution disease victim by the city did not help their lives at all.

On June 10, 1966, Kihira Usaburo, an old and frail asthma patient, hanged himself. Next to his body was a suicide note saying that he wanted to live comfortably in the afterlife where he didn't need medicine or money. At the funeral, the representatives of the Pollution Disease Patient Protection Association staged a protest, asserting four desperate demands to the city authorities. These were welfare for pollution disease patients, the establishment of an emergency care center near the mountain, factory fuel replacement, and a halt to industrial expansion. The city authorities, instead of responding to this, were only focused on approving the third complex

[1] https://en.wikipedia.org/wiki/yokkaichi_asthma

plan. Eventually, in February 1967, despite strong opposition from the residents, the city council achieved its goal through a coercive vote.

On June 13, 1967, the second suicide incident occurred. Katsuhiro Otani, who had held the portrait of the first suicide victim with a black band in front of the funeral procession, took his own life. Otani's lungs could no longer withstand the gray dust that filled Yokkaichi. This was written in a diary left by Otani.[2]

> On June 3, 1966, the smog got worse after 5 p.m. I couldn't stop coughing. I packed my lunch and hurriedly left the village. It was utterly miserable. I wanted to stay home, but I had to come to this desolate place. Why doesn't the damn mayor get asthma? Even if he gets sick, would he understand our feelings? Ah, I don't want to die because of this dirty air.

Otani's death drove asthma patients to extreme conditions. There had been talk of suing when Kihira died, but it was just talk at the time. However, the attitudes of asthma patients noticeably changed when Otani, who was active and respected by those around him, died. On September 1, 1967, nine Yokkaichi asthma patient groups filed a damage lawsuit against six companies for their responsibility for the pollution. On July 24, 1972, five years after the legal battle began, the court ordered six companies within the petrochemical complex to pay a total of 88 million yen in damages to the plaintiffs. It was very innovative at the time in Japan to make a corporate group responsible for compensation. Everyone in the nearby area suffered from the Yokkaichi incident, and a total of 1,231 people officially recognized as having suffered serious damage, of which more than 80 had already died. The Japanese government knew in 1961 that sulfur dioxide was causing serious problems, but it wasn't until February 1969, after this incident, that they finally established emission standards.

The Japan Environmental Agency, established in July 1971, began to strengthen regulations after reviewing the status of air pollution regulations. They redefined the standards for sulfur dioxide and created a law to regulate pollution based on total amount rather than concentration, and encouraged the use of low-sulfur petroleum. The Yokkaichi ruling reminded people that the size of industrial complexes should be limited in consideration of local environmental capacity, along with increased interest in pollution control in the industrial sector. Also, companies learned through this case that they could be sued if the pollution from multiple factories within the complex, even if each factory meets the air pollution standard, combines to harm local residents.

The Yokkaichi incident caused a huge ripple in Japan, which was devoted to rapid economic growth after the war. The early 1960s was the peak of Japan's economic expansion, and all local governments across the country were eager to establish complexes. However, as a result of this incident, movements opposing the establishment of complexes began in some areas, and more careful reviews began. Also, this incident had a significant impact on the establishment of Japan's Environmental Agency in 1971 and air pollution regulations. Especially, it played a decisive role in the enactment of the 'Health Damage Compensation Act due to Pollution' in 1974,

---

[2] Island of Dreams: Environmental Crisis in Japan, Norrie Huddle and Michael Reich, Schenkman Books, 1987.

which mainly compensates victims for medical expenses and living expenses caused by the polluter.

Chapter 8
Mexico City Air Pollution

In February 1987, a shocking incident occurred in Mexico City, the capital of Mexico, where thousands of birds flying in the sky fell and died. Upon examining the bird carcasses, large amounts of lead, cadmium, and mercury were detected in the heart, lungs, and liver. These substances were found to have concentrated in the birds' bodies due to severe air pollution in Mexico City.

Mexico City is located on a plateau at an altitude of 2,240 m, surrounded by mountains 4,000–5,000 m high. About 20 million people lived in an area of 4,902 km^2, and about 5 million cars were running around. Moreover, at that time, there were about 35,000 factories scattered throughout the city, including oil refineries and thermal power plants. These factories burned about 45 million tons of fossil fuels in a single day, and intense sunlight and dry climate added to this, causing photochemical smog all year round. Also, a large amount of lead polluted the air because leaded gasoline was used as car fuel until the mid-1990s.

According to a survey by the World Health Organization (WHO), the air pollution in Mexico City in the 1980s was the worst among major cities worldwide.[1] Since 1986, the WHO began measuring six air pollutants, including sulfur dioxide, carbon monoxide, lead, nitrogen oxides, particulate matter, and ozone, at 28 locations in Mexico City. All of the measurements exceeded each criterion three to four times.

Citizens of Mexico City, who had to breathe such air, suffered from symptoms such as headaches, insomnia, lethargy, and vomiting, and in severe cases, they even reported hallucinations and delusions. In fact, the concentration of heavy metals in the bodies of local residents was more than 20 times higher than in Germany or the United States. At that time, experts in Mexico City warned not to breastfeed young children because a large amount of heavy metals was contained in breast milk. This is because heavy metals accumulate more in breast milk, which is high in fat, and the concentration is much higher than in blood or body fluids. Not only that, but

---

[1] https://en.wikipedia.org/wiki/air_pollution_in_mexico_city.

S. S. Park, *Eco-disasters*, https://doi.org/10.1007/978-981-96-8667-4_8

the property damage caused by air pollutants, such as corrosion of buildings and vehicles, was also very serious.

In Mexico City, expatriate employees working for foreign companies received special hazard allowances due to the severe air pollution, and travelers were warned not to open windows in their hotel rooms for ventilation. Doctors declared that living in Mexico City at the time was as harmful as smoking more than two packs of cigarettes a day. There are several reasons why Mexico City has the worst air pollution in the world. Among them, the city is located in a high-altitude basin, which hinders air mixing and has 23% less oxygen than flatlands. The city, with a population of over 20 million, has expanded unplanned in this high-altitude area, and the main factors include the emission of smoke from factories scattered within the city and numerous cars. Mexico City is surrounded by high mountains on three sides, and cold air rushes in through the open space on the remaining side, pushing up the hot air.[2] The hot air layer in the upper atmosphere becomes a black cloud and acts as a lid. At this time, the cold air on the ground is trapped in the basin terrain, causing smog to linger near the ground, and the temperature inversion phenomenon repeats every day. Also, due to economic circumstances, lead-containing gasoline harmful to the human body was used in vehicles, and due to the dry climate, a lot of dust containing combustion gas and heavy metals was not washed away by rainwater, but went up and down between the ground and the atmosphere, exacerbating air pollution.

In Mexico City, politicians have been promising to solve the air pollution problem for decades, but those promises have always ended without any action due to difficult economic conditions. Ironically, the air pollution problem that was neglected due to economic excuses came back as serious damage to the national economy in the mid-1980s. Respiratory diseases became the second leading cause of death in Mexico City after heart disease, and the medical expenses that the country had to pay for respiratory diseases doubled every three years. The labor efficiency of workers in Mexico City was halved compared to other regions in Mexico, and the learning ability of young children was also found to be 25% behind. Also, as Mexico City's air pollution became widely known to the international community, the number of tourists and foreign investors rapidly decreased.

Since the early 1990s, a large-scale air pollution prevention plan has been launched in Mexico City. In the first five years, 47 billion dollars were invested and 112 projects were implemented, including low-pollution fuel replacement, strengthening emission regulations, relocating polluting industries, restoring vegetation in wastelands, and establishing car-free days. Mexico City has gradually improved the quality of all fuels used and modernized public transportation, so that all The obligation to attach state-of-the-art combustion catalysts to vehicles and strengthen emission regulations was enforced. The use of leaded gasoline was gradually reduced until it was completely discontinued in 1997 and replaced with unleaded gasoline. Also, from November

[2] Cluster Analysis of the Wind Events and Seasonal Wind Circulation Patterns in the Mexico City Region, Susana Carreón-Sierra et al. (2015). https://www.mdpi.com/2073-4433/6/8/1006.

1989, a five-day rotation system was implemented for all vehicles, excluding diplomatic vehicles. Mexico City has continuously modernized its lagging industries and strengthened emission regulations, designating 58% of the total city area as environmental land to limit development and restore ecosystems, and fundamentally re-adjusted development policies by planting trees in the remaining urban areas. They also conducted proactive environmental education and fostered programs that citizens could voluntarily participate in. In 1994, Mexico became the first Central and South American country to join the Organisation for Economic Co-operation and Development (OECD) and started a large-scale research project with the help of international environmental experts. In the late 1990s, for international promotion, they also involved Mario Molina, a Mexico-born professor at the Massachusetts Institute of Technology (MIT) in the United States, who won the 1995 Nobel Prize in Chemistry for research on how Freon gas destroys the ozone layer.

As a result of these efforts, the air in Mexico City has greatly improved over the past 10 years. Carbon monoxide, sulfur dioxide, and ozone concentrations have decreased to one-third, and nitrogen oxides and fine dust have decreased to half. The lead content in the air also dropped to less than one-tenth. However, due to unavoidable geographical and climatic conditions, Mexico City remains one of the world's largest cities where air environment management is very difficult.

Part III
Poisoned Water and Food

Chapter 9
Minamata Disease

In 1952, a strange disease broke out in a small fishing village called Minamata in Kyushu, Japan.[1] Water birds flying in the sky suddenly fell to the ground, and cats kept at home started to spin around like crazy, foaming at the mouth. This bizarre phenomenon appeared in this village. At first, the villagers called it 'dancing cat disease'. Then, a year later, this disease also broke out among the residents here. The main symptoms were paralysis of the hands and feet, pain and chills, headaches, visual impairment, motor disorders, and language disorders. In severe cases, intense pain and paralysis symptoms led to death, and babies born were either deformed or stillborn.

Initially, health authorities and doctors could not find the cause of this disease, which had never been reported worldwide, and thought it was a simple endemic disease. However, as the number of people with this disease increased and the symptoms continued to appear in cats and water birds, in 1956, Minamata City formed a countermeasure committee to investigate the cause of the disease. However, the committee made a wrong diagnosis that this disease was a type of Japanese encephalitis and could not establish any special measures.

In July 1959, Professor Takeuchi Tadao of Kumamoto University claimed that such symptoms occur when there is some damage to the central nervous system and the brain, and it is due to mercury poisoning, a liquid heavy metal.[2] It was later revealed that the seawater and seabed sediments of Minamata Bay, and the fish caught nearby, contained a large amount of mercury, and the consumption of such fish was the direct cause of these symptoms. It was also concluded that the reason why Minamata Bay was contaminated with mercury was due to the wastewater discharged from the adjacent New Japan Chisso Fertilizer Factory.

The mercury contained in the wastewater discharged from the New Japan Chisso Fertilizer Factory was diluted in the sea, but it was transformed into a more toxic

[1] https://minamata195651.jp/pdf/kyoukun_en/kyoukun_eng_all.pdf.

[2] Island of Dreams: Environmental Crisis in Japan, Norrie Huddle and Michael Reich, Schenkman Books, 1987.

S. S. Park, *Eco-disasters*, https://doi.org/10.1007/978-981-96-8667-4_9

substance called methylmercury by microorganisms, concentrated in the bodies of the creatures living here, and the concentration increased as it moved up the food chain. Therefore, the toxicity was greatly manifested in water birds, cats, and people who had been eating fish with a high concentration of mercury for a long time. This is an example of biomagnification, a phenomenon where the concentration of harmful substances accumulated in the surrounding environment or lower organisms in the food chain increases as they move to higher levels of the food chain. For example, even if mercury exists in seawater at a very low concentration, it can accumulate in fish bodies at a concentration thousands or tens of thousands of times higher. At the time, the mercury concentration in the seawater of Minamata Bay was up to 0.0006 ppm, but the fish were concentrated up to 80,000 times at 10~50 ppm.

Kumamoto Prefecture and the Japanese central government acknowledged the problem in Minamata Bay at the time, but did not acknowledge the cause. Kumamoto Prefecture tried to declare this area a fishing prohibition zone under the Food Sanitation Act, but the central government continued to allow fishing activities on the grounds that there was no clear evidence that the entire coastal waters were contaminated. Also, the central government allowed the discharge of factory wastewater, saying that while it was certain that there was a problem with the factory wastewater, it was impossible to prohibit discharge unless the causative substance was clearly identified. It was an absurd decision.

Eventually, the factory continued to operate and some residents began to leave Minamata. However, due to the continued struggle of the remaining residents, the factory agreed in December 1959 to pay 300,000 yen to the deceased, 100,000 yen per year to adult patients, 30,000 yen per year to child patients, and 20,000 yen for funeral expenses as consolation money, and this incident was concluded. However, this consolation money was different from compensation, meaning that the factory side would not admit their responsibility.

Six years later, in 1965, residents living downstream of the Agano River in Niigata Prefecture also reported symptoms similar to Minamata disease. The residents definitively identified the wastewater from Showa Denko, located upstream, as the cause and filed a lawsuit. This was the first environmental lawsuit to go to court in Japan. Subsequently, the Yokkaichi incident lawsuit in 1967 and the Itai-itai incident lawsuit in 1968 were filed one after another. In September 1968, the Japanese Ministry of Health and Welfare officially recognized that the causative substance of the Niigata incident was an organic mercury compound. As a result, the Japanese government, which had not recognized the causative substance of the Minamata incident for 15 years, announced that it was mercury from the Chisso factory wastewater.

Then, in 1969, the citizens of Minamata filed a lawsuit demanding 640 million yen in compensation, nullifying the agreement with the Chisso factory made 10 years ago in 1959. After that, the victims' sit-ins and legal battles continued, and finally, in 1973, the victims won. The judge in charge of the case acknowledged in the judgment that the factory side had committed a crime of thoroughly concealing the discharge of mercury into Minamata Bay, and declared a plaintiff's victory in all respects. Also, this judgment ordered the payment of 937 million yen in compensation to the 30 families who filed the lawsuit. The factory side They agreed that the wastewater caused Minamata disease and officially acknowledged their responsibility. At the time, the damage status showed that 43 people died and 111 showed symptoms of incurable paralysis, and 19 children were born with serious deformities.

Meanwhile, another problem arose around that time. The number of victims who were not recognized as patients increased every year, and they also went to court. Moreover, not only in Minamata and Niigata, but also in Niigata, Tokyo, Kyoto, Osaka, and Fukuoka, patients complaining of symptoms similar to Minamata disease continued to appear and began to participate in legal battles.

The legal battle of unrecognized patients continued for decades and finally ended on May 19, 1996. A total of 1,984 patients filed lawsuits, and they finally agreed on a compensation of 2.6 million yen per person and withdrew the lawsuit. The reason they agreed to such a small amount of compensation was because the number of survivors who should receive compensation was decreasing and the perpetrator's ability to compensate had reached its limit. Therefore, the Japanese government chose a method of creating a fund, lending it to the perpetrator, and using that money to pay compensation to the victims.

New Japan Chisso Fertilizer, the perpetrator of the incident, was a leading figure in Japan's modernization, starting to produce fertilizers, explosives, and chemical products after establishing a Minamata factory in 1908. However, since the Minamata disease surfaced in the early 1950s, the company's suffering from this issue continued for decades. Moreover, the victims suffered the lifelong pain of physical ailments that were incomparable to material damage.

People in Minamata appealed to the government not to attach their hometown name to this terrible disease anymore. Heavy metals, radioactive substances, and toxic chemicals can cause mutations in human genes, so even if symptoms do not appear in the present generation, they can be passed on to descendants. For this reason, those from Minamata had to suffer the pain of being disadvantaged in various aspects such as marriage and employment.

The Minamata incident took over 40 years from occurrence to conclusion, making it the longest-lasting environmental incident of the twentieth century. Agreements on compensation between the perpetrator and the victims were made three times in 1959, 1973, and 1996. However, it is estimated that there are still over 10,000 people in Japan who want to be recognized as Minamata disease patients. It is an exemplary case that showed the world the horror and tediousness of diseases caused by pollution.

The United Nations Environment Programme (UNEP) signed the 'Minamata Convention on Mercury' in Kumamoto, Japan on October 9, 2013, to reduce the harm to humans and the environment caused by mercury, and it came into effect in 2017. Because mercury has the characteristic of long-distance travel in a gaseous state, the need for international joint response to the entire process from production, use, emission, and disposal has been raised for a long time, which led to an international convention. From 2020, the convention aims to limit the trade of mercury raw materials and gradually prohibit the manufacture and import/export of mercury-added products.

Chapter 10
Hinkley Chromium

In 1999, a movie called "Erin Brockovich," which dealt with a real-life environmental lawsuit in the United States, was released, causing a nationwide stir. This movie dramatized the truth-finding and legal battle process of the 'Hinkley Chromium Case', which occurred due to groundwater pollution in California from 1993 to 1996. The movie was successful at the box office, stimulating the interest of environmentally conscious Americans, and was also recognized for its excellence, being nominated for the Academy Awards in 2001. The film was made more interesting by the casting of Julia Roberts, known as the actress with the most beautiful smile in America, a far cry from the depressing and cruel environmental incidents. She won the Academy Award for Best Actress for this film in 2001.

The beginning of the Hinkley Chromium case dates back several decades.[1] In 1952, a company called PG&E (Pacific Gas and Electric Company), headquartered in San Francisco, USA, and supplying electricity and gas to the western region, set up a gas compressor station in a place called Hinkley. Hinkley is a small town located in the Mojave Desert, halfway between Los Angeles, California, and Las Vegas, Nevada. The compressor station built here was to pressurize natural gas produced in the southern region of California and send it to the central and northern regions of California. To cool the heat generated in the process of gas compression, PG&E installed a cooling tower and added a heavy metal called Hexavalent Chromium to the cooling water to prevent corrosion of the cooling system. The cooling water and various wastewater from the compressor station were dumped into a nearby pond.

As time passed, the residents of Hinkley began to suffer from various diseases, including cancer, in the liver, heart, brain, kidneys, respiratory system, and reproductive organs. Livestock raised in the ranches and barns of this town also died of strange diseases, and all animals, including frogs, were found dead in the pond where the wastewater was dumped. However, no one paid attention to the strange

[1] https://en.wikipedia.org/wiki/Hinkley_groundwater_contamination.

S. S. Park, *Eco-disasters*, https://doi.org/10.1007/978-981-96-8667-4_10

things happening in this small rural town, and the residents lived for decades, considering it only a strange phenomenon. Then, the truth of this case was revealed to the world through the efforts of a curious female employee named Erin Brockovich, who worked at a law firm called 'Masry & Vititoe'.

This incident is just a common type of environmental disaster where harmful heavy metals contaminate groundwater and residents suffer from various diseases. However, this incident drew significant attention from Americans because the truth was revealed through the tenacity of a woman named Erin Brockovich, who had a unique career and personality, and because a large corporation had long known the truth but tried to cover it up with hypocrisy and lies. The incident caused even greater repercussions when it was made into a movie. As soon as the movie was released, a situation arose where hexavalent chromium was investigated throughout the United States. Regardless of rural areas, small and medium-sized cities, or large cities, groundwater, rivers, lakes, and other places where harmful heavy metals could be found were all investigated, and the analysis results were made public in the media.

Erin Brockovich, who led the Hinkley Chromium case, was a woman who couldn't study hard due to congenital dyslexia. She had no decent education, no qualifications or talents. On top of that, she had been divorced twice and had to raise three children, and she was in a desperate situation without a job or money. She tried to find a job and ran around in all directions, but it was all in vain. Eventually, she had a traffic accident in Reno, Nevada in 1991 and went to a lawyer named Edward Masley, whom she met through the accident, to beg for a job. Kind-hearted lawyer Masley, hearing her desperate situation, reluctantly gave her a job handling paperwork and odd jobs. The two, having met in this way, later became overnight stars that shook the entire United States and were fortunate to receive millions of dollars in case fees.

The circumstances of the case are as follows. One day, Erin came across a document she couldn't understand while organizing files in the lawyer's office. She found it very suspicious that a mysterious medical diagnosis was attached to a file classified as a real estate-related document. Unable to contain her curiosity, Erin went directly to Hinkley to examine the whole story and realized that this was not a simple real estate transaction. Most of the villagers were suffering or dying from similar diseases, and strangely, a company called PE&G was paying for all the villagers' medical expenses. The medical diagnoses she saw in the lawyer's office were attached for precisely this reason. Moreover, while the villagers were suffering from various diseases, the company was buying land and houses at prices higher than the market price. It was hard to accept the fact that they were buying at a higher price when they could buy flawed real estate at a lower price. Erin immediately informed lawyer Masry of this fact and began a full-scale investigation.

During her investigation, Erin discovered that the various diseases that appeared among the residents were due to the contamination of Hinkley's soil and groundwater by wastewater discarded from the gas compressor station. She found that the wastewater contained a highly toxic heavy metal called hexavalent chromium, and was certain that the residents had contracted diseases by using the contaminated groundwater as drinking water. It was obvious that the residents would have been living by inhaling steam containing harmful heavy metals while showering with this water and using it as cooling water for air conditioners. Moreover, children always played with sprinklers in the hot desert and spent their days in swimming pools. The residents had been drinking and breathing in toxic heavy metals, and absorbing them into their bodies through their skin.

Erin discovered that the damage and toxicity of hexavalent chromium to the human body matched the diseases that the residents of Hinkley were experiencing. Erin began to persuade the residents and lawyer Masry to sue PG&E. However, the townspeople, who had always thought that all diseases were their own fault, did not readily agree with Erin's opinion. Masry also wanted to end his role with just real estate transactions. Although everyone around her discouraged her, saying it was a fight she could not win, she eventually started the largest environmental lawsuit in American history.

As the lawsuit progressed, astonishing facts were revealed. PG&E had known since 1965 that the corrosion inhibitor they used contaminated Hinkley's soil and groundwater. However, the company ignored this and illegally discharged a huge amount of corrosion inhibitor into a nearby pond. The toxicity of hexavalent chromium, included in the corrosion inhibitor, had been known since around 1920, but this fact was ignored, and about 370 million gallons (1.46 billion liters) of wastewater had been dumped here. PG&E knew in 1987 that the groundwater in the Hinkley area was contaminated 10 times higher than the environmental standard.

When it became known that wastewater from the compressor station had contaminated Hinkley's groundwater with the heavy metal chromium, the company hid the fact that chromium can be toxic depending on its form and began spreading false information to the residents, making it seem beneficial to health. Moreover, PG&E dismissed residents' complaints by claiming that the groundwater quality was very good and there was no problem as drinking water.

Chromium is a very unique heavy metal, its toxicity varies greatly depending on its form. For example, trivalent chromium is not only harmless to the human body, but is also an essential nutrient. It is involved in the metabolism of sugars and fats, and is also essential for insulin function. If there is a lack of chromium, symptoms similar to diabetes appear and it is related to arteriosclerosis and high blood cholesterol levels. However, hexavalent chromium is toxic to the human body. It can cause lung cancer, bronchial cancer, kidney disorders, and can cause dermatitis and ulcers upon skin contact. The damage is particularly severe when inhaled through the respiratory system. Hexavalent chromium is used as a pigment, dye, disinfectant, and anti-corrosive, and plating, leather, dyeing, alloy, and cement factories are the main sources. Human damage is mainly caused by workers in plating factories who inhale chromium in a vapor state, often causing lung cancer or skin diseases.

PG&E cunningly exploited these unique properties of chromium to deceive the residents. As the four-year court battle progressed, the truth gradually emerged. Erin found scientific evidence while attending college and the waterworks office. She also met with people who had worked at the former compressor station and residents in turn, digging up the secrets hidden there. Initially indifferent residents became more actively involved as the case progressed. The number of plaintiffs, which was 77 when the lawsuit began in 1993, increased to 634. In the end, the court ordered

PG&E to compensate the Hinkley residents with a huge sum of $333 million, clean up the polluted area, and stop using the problematic hexavalent chromium. In 2000, a large corporation with a capital of $35.29 billion and an annual sales of $26.23 billion bowed to the tenacity of a seemingly insignificant woman.

PG&E suffered enormous losses from this case, including damage to the company's image and asset losses due to compensation. After the incident, the giant corporation PG&E faced bankruptcy in 2002. Despite the common belief that utilities such as electricity and gas supply are structurally indestructible, PG&E was an exception for disregarding environmental ethics. PG&E's bankruptcy is a good example showing that even a giant corporation standing on a rock can collapse if it disregards the environment.

Chapter 11
Itai-Itai Disease

Residents living in the Jinzu River basin in Toyama Prefecture, Japan, began to experience severe pain in their waist and joints around 1910. The longer a person had lived in the area, the more severe the symptoms were, with frequent fractures of the arms, ribs, pelvis, and femur. The bones became so weak that even a cough could cause a fracture, and due to bone atrophy, height decreased, with severe cases seeing a reduction of more than 20 cm.

These symptoms were particularly prevalent among farmers living near the river and their families. Those who did a lot of labor, had lived in the area for more than 20 years, or women who had given birth, experienced more severe symptoms. However, at the time, it was considered a simple regional endemic disease, and decades passed without even knowing the cause or the name of the disease. The disease was only called "Itai-itai" disease, which means "it hurts, it hurts" in Japanese, because those who contracted the disease complained of severe pain.[1]

First attempts to identify the cause of the disease began shortly after the end of World War II.[2] The diagnosis was vitamin D deficiency. Therefore, vitamin D was distributed to local residents for treatment, but it had no effect. Eventually, it turned out to be a misdiagnosis.

Around this time, a doctor named Hagiwara Noboru, who lived in the area, took a special interest in the disease. He investigated the symptoms and number of deaths and reported to the academic community in 1955. In 1961, he and Professor Kobayashi Jun of Okayama University identified the cause of the disease. They concluded that the Itai-itai disease was caused by cadmium discharged from the Mitsui Metal Mining in Kamioka, 50 km upstream of the Jinzu River. They detected high concentrations of lead and zinc in the patient's body tissues, and particularly confirmed cadmium reaching thousands of ppm. There were even patients with cadmium concentrations of 10,000 ppm. Mitsui Mining had been producing copper, zinc, and lead there for

---

[1] https://en.wikipedia.org/wiki/Itai-itai_disease.

[2] Island of Dreams: Environmental Crisis in Japan, Norrie Huddle and Michael Reich, Schenkman Books, 1987.

S. S. Park, *Eco-disasters*, https://doi.org/10.1007/978-981-96-8667-4_11

over half a century. However, the cadmium in the mine's waste ore contaminated the Jinzu River, and this river water flowed downstream and was used as agricultural water for paddy fields and fields and as drinking water for residents, thus introducing cadmium into the human body. Cadmium, which has a similar structure to calcium, accumulates in the bones in the animal body and causes disorders in the liver, kidneys, and thyroid function.

However, Mitsui Mining firmly denied this, claiming that the cause of Itai-itai disease was a lack of vitamin D, as had been concluded in the past. The local medical association also supported the claim of Mitsui Mining. They distributed vitamin D, which was again lacking, to the residents and made them consume it, but it had no effect. The Japanese Ministry of Health and Welfare finally acknowledged in 1968 that the cadmium from Mitsui Mining was the cause of the disease. At that time, 56 people had died from this disease, and hundreds had suffered. However, it is obvious that there would have been many more victims of cadmium discharged from this mine over more than half a century.

When the Japanese Ministry of Health and Welfare recognized cadmium as the cause of the disease, the patients sued Mitsui Mining, which had denied it for seven years. This became one of the four major environmental lawsuits in Japan at the time, along with Niigata, Minamata, and Yokkaichi. However, unlike other cases, the victims of Itai-itai disease did not need to fight a long legal battle against a company that refused to take responsibility. This is because the Japanese Mining Law, enacted in 1939, strictly enforced systems like no-fault compensation insurance to prepare for potential risks from mining operations. This law stipulated that compensation would be made if the cause was proven. On June 30, 1971, the judgment handed down in the Toyama court gave the plaintiff a complete victory. Moreover, this judgment acknowledged that epidemiological evidence of the cause is sufficient legal evidence if it is proven without contradiction pathologically and clinically. This was used as a useful precedent in other environmental lawsuits later.

Cadmium is highly toxic and even a small amount entering the body can cause vomiting, diarrhea, abdominal pain, gastritis, headaches, and muscle pain. Acute poisoning at high concentrations can cause respiratory distress, chest compression, loss of appetite, heart and lung failure, and in severe cases, can lead to death. Chronic poisoning at low concentrations can cause rhinitis and olfactory disorders, headaches, insomnia, anemia, liver dysfunction, kidney dysfunction, and osteomalacia. Generally, when heavy metals enter the body, they accumulate in the body over a long period of time, and in the case of cadmium, it is known that the residence time in the body reaches 16–33 years.

As the damage caused by cadmium became official, the Japanese government began to establish measures to prevent cadmium pollution in May 1968, such as setting the allowable concentration of cadmium in water and rice, and setting standards for human intake and industrial wastewater discharge. However, there were difficulties in applying these regulations. For example, in the case of rice, the cadmium concentration before milling (the process of removing rice bran from the rice mill) was initially determined to be less than 0.4 ppm, but in July 1970, the Food Sanitation Act was amended to change the standard to 1 ppm for rice before milling and 0.9 ppm after milling. In fact, there was almost no rice that could be sold according to the initial standard, so the regulations had to be relaxed.

In 1971, the newly established Japanese Environmental Agency began investigating farmland that produced rice contaminated with cadmium. In 1972, the Environmental Agency investigated 117 places equivalent to a total area of 11,700 ha and revealed that 25% of them produced rice contaminated with a cadmium concentration of 1 ppm or more, and soon issued a production ban. The rice produced from 25% of the farmland investigated did not even meet the relaxed standards.

The Itai-itai incident was another environmental disaster caused by mine wastewater that Japan experienced following the Ashio mine incident at the end of the nineteenth century. From the late nineteenth century to the early twentieth century, Japan mined lead, zinc, and copper in many places for about 100 years in preparation for war, but the wastewater and waste from mines and refineries were simply discarded in nature. After the war ended, as the mines became unprofitable, they were abandoned without any further management. Also, with the development of the electrical and electronic industries after the war, the use of cadmium increased rapidly, and the pollution that was once limited to rural areas around mines spread to urban areas. In particular, cadmium was used in electronic components, alloys, electroplating, dyeing, ceramics, photography, reactors, etc., so the sources of pollution are scattered in many places. Cadmium is also emitted when plastic products are incinerated at waste incineration plants. The impact of cadmium pollution that started with the Itai-itai incident swept across Japan along with rapid industrialization after the war.

Chapter 12
Milwaukee Cryptosporidium Outbreak

On April 3, 1993, the emergency room of a hospital in Milwaukee, Wisconsin, was filled with patients complaining of diarrhea. Many students and citizens were taken to the hospital emergency room instead of going to school or work. The city of Milwaukee was in a state of shortage of diarrhea medication due to the countless patients pouring in. Only then was it revealed that a waterborne epidemic had occurred in the city. Initially, the authorities, who could not identify the pathogen and cause of the disease, could only recommend measures such as boiling all drinking water.

Two weeks after the outbreak, the cause of the disease was found to be tap water supplied by the Milwaukee Southern Waterworks, and the pathogen was Cryptosporidium. The contaminated tap water was supplied to citizens from March 23 to April 8, when the water supply was stopped. During this period, the turbidity of the tap water was recorded as abnormally high due to the increased number of Cryptosporidium.

It was estimated that about 52% of the citizens who drank the tap water were infected with this pathogen. Some citizens only showed mild symptoms such as loss of appetite, but about 403,000 people showed serious symptoms such as abdominal pain, diarrhea, and vomiting, of which about 4,400 were hospitalized and 109 died.[1] In addition to this enormous human damage, it was estimated that there was an economic loss of $1.7 million in medical expenses and $64.6 million due to decreased productivity. This incident was recorded as the worst tap water accident in the developed world in the late twentieth century.

Milwaukee supplies tap water from Lake Michigan, and it undergoes standard water treatment consisting of coagulation, sedimentation, filtration, and disinfection. However, just before the accident, heavy rain and melting snow caused high turbidity river water to flow into the water intake point. In addition, high concentrations of Cryptosporidium were discharged from slaughterhouses, sewage treatment plants, and ranches located upstream of the river, but they were not completely removed

[1] https://en.wikipedia.org/wiki/1993_Milwaukee_cryptosporidiosis_outbreak.

S. S. Park, *Eco-disasters*, https://doi.org/10.1007/978-981-96-8667-4_12

during the water treatment process. Despite these problems in the source water and the treatment process, the tap water met all standards, and especially passed the test for coliform bacteria, an indicator of pathogenic microorganisms. Moreover, turbidity, which was abnormally high due to cryptosporidium, was also passed without any problems in the tap water purification standard.

Cryptosporidium, the cause of the incident, is a protozoan that causes diarrhea. In a way, it is a factor in the outbreak of waterborne diseases because it has strong resistance to chlorine disinfection and environmental resistance. Protozoa are single-celled animals that are larger than bacteria, but can only be observed through a microscope. Cryptosporidium typically inhabits the intestines of animals and is about 1/16 the size of dust particles floating in the air. The main sources of contamination are human, livestock, or wildlife feces, and infection occurs through food or drinking water, with the most common route being transmission between humans or between humans and animals. Infections through tap water are rare and unlikely, but once an infection occurs, it is widespread and has a large ripple effect, requiring careful management.

This disease, medically called Cryptosporidiosis, usually heals naturally after about a week in healthy people, but in people with weak immunity, it does not recover and leads to chronic diarrhea. In patients with immune system problems, including AIDS, they experience rapid weight loss due to chronic diarrhea, eventually becoming weak and leading to death. About half of the deaths in the Milwaukee incident were AIDS patients with immune deficiencies. At that time, it was estimated that 20–30% of AIDS patients in Milwaukee died from this disease. Since no special treatment has been developed for this disease other than natural healing, special attention is required for the elderly and people with weak immunity, such as AIDS patients.

Cryptosporidium is distributed at low concentrations in surface water such as rivers and lakes. Depending on the region, this protozoan exists in 65~97% of surface water, and there are reports that it is detected in the feces of 1~4% of patients with gastrointestinal diseases. This means that you can easily get infected with this disease when you use surface water for drinking water. Also, protozoa have a spore formation period when a hard shell is formed, so it is impossible to completely remove them with general chlorine disinfection. To treat this, it is known that strengthening the sand filtration process, ozone disinfection, and long-term high-concentration chlorine disinfection are effective. Even drinking spring water needs thorough management because it can easily be exposed to this disease, and especially people with weak immunity should always boil the water from rivers or lakes when camping and refrain from water play or swimming.

In 1984 in Texas, USA, 1987 in Georgia, 1996 in Saitama Prefecture, Japan, 1998 in Lake Sydney, etc., Cryptosporidium was also found through drinking tap water or swimming pools. There have been cases of similar infections. Such accidents are occurring every year not only in the United States but also in developed countries such as the United Kingdom, Australia, and Japan, and it is reported that more than 500,000 people have been infected over the past 10 years. Particularly, many people were infected in group incidents in Milwaukee and Georgia (1,700 people infected in 1987), both of which were properly handled and caused a great shock as they occurred in tap water disinfected with chlorine. This means that when using surface water contaminated with protozoa as raw water, methods other than chlorine disinfection are needed. Giardia, the most common cause of diarrhea worldwide, is a protozoan that shows strong resistance to chlorine disinfection and is mainly infected through tap water.

In Missouri, USA, frozen water pipes burst, causing sewage to enter the drinking water in the winter of 1989. This incident led to 240 people being infected with *E. coli* (O157:H7), resulting in 32 hospitalizations and 4 deaths. In 1991, 59 people were infected with *E. coli* after swimming in a lake in Portland, Oregon. In 1998, similar symptoms appeared in 26 children at a swimming pool near Atlanta, Georgia. In 1990, a water tank at a hospital in Chicago, USA, was contaminated, and 21 people were infected with a protozoan called Cyclospora, which causes diarrhea. In 1993, more than 600 people in Missouri were infected with a bacteria called Salmonella, 15 of whom were hospitalized and 7 of whom died. This was found to be due to the infiltration of wild bird feces into the water tank (reservoir) installed to drain the water produced at the water purification plant.

The Milwaukee incident was an important opportunity for the U.S. government to improve water purification technology and management systems. In 1996, the U.S. Congress amended the 'Drinking Water Management Act' enacted in 1974 to strengthen the management and surveillance system for tap water. Therefore, from December 1998, more thorough management of Cryptosporidium was started for drinking water. Also, when using surface water as the source of tap water, continuous monitoring of protozoa should be carried out, and new water treatment methods should be applied where problems may occur. In particular, it was decided to expand the use of riverbank filtered water and subsurface filtered water by digging wells on the riverbank or bottom instead of directly drawing water from rivers or lakes for purification. Also, long-term care facilities and medical facilities were required to establish a surveillance system directly linked to a water quality inspection office. In Milwaukee, water treatment was carried out according to the existing regulations of the state and federal governments. Despite the occurrence of a waterborne disease outbreak, this incident served as a turning point, reminding us of the need for improved cooperation between disease surveillance systems and public health authorities.

In the United States, the Milwaukee incident led to strengthened laws and regulations and advanced water purification technology, greatly improving the quality of tap water. However, this incident amplified Americans' distrust in tap water, leading to a surge in the use of bottled water. In the United States, it is legal to purify tap water and sell it in plastic bottles, not just natural spring water. Before the Milwaukee incident, it was limited to natural spring water (Spring Water, Mineral Water), but after the incident, distrust increased, leading to a continuous increase in the sales of products that re-purify tap water and bottle it (such as Coca-Cola's DASANI, Pepsi-Cola's AQUAFINA, etc.), now occupying half of the market.

The supply of tap water has greatly contributed to improving human health and extending life over the past 100 years. In particular, the chlorination of tap water, which began in the United States in 1908, has saved countless people from waterborne diseases. However, in developing countries that do not receive tap water, waterborne diseases are still one of the most terrifying diseases threatening human health. The Milwaukee incident showed that the water problem is not limited to developing countries. In fact, in the United States alone, the population drinking water that does not meet the federal environmental protection agency's water quality standards reaches 6 million, and every year 1 million people suffer from waterborne diseases, with about 900 reported deaths. If bacteria and viruses in tap water are completely removed by chlorination and a certain amount of residual chlorine is maintained during the tap water supply process, the safety of tap water can be guaranteed. However, in the case of protozoa, the safety cannot be guaranteed with the current tap water treatment and supply process we are implementing. This requires more thorough raw water management and surveillance of the purification and supply process. Also, waterborne diseases are often infected not only through drinking water but also through swimming pools, bathhouses, and outdoor water play.

Since waterborne diseases often recover on their own through the body's immunity, the cause is often overlooked. This is well illustrated by the fact that in the health statistics of developed countries, the proportion of diarrheal diseases with unknown causes is more than 30%. This phenomenon means that we can easily be exposed to waterborne diseases and can be infected without knowing it. And it suggests that water is threatening our lives on a daily basis. To protect our lives from this threat of water, we need to focus on water source management, purification treatment, supply process, and use. Even more thorough management and surveillance are necessary until then, and at the same time, the introduction of technology and systems that can support this is required.

Chapter 13
Niigata Mercury Poisoning

In 1964, mysterious disease of unknown cause appeared among the residents living in the Agano River basin of Niigata Prefecture, Japan. Patients suddenly felt numbness and paralysis in their arms and legs, unable to walk properly, and even unable to use chopsticks.[1] They also developed language disorders and body convulsions, and in severe cases, they died. Deformed babies who could neither speak nor crawl were born from mothers without symptoms. Most of the patients were fishermen who often ate fish while living by the river, and during the examination process, it was revealed that their symptoms were very similar to Minamata disease.

Niigata Medical University diagnosed that this disease may have originated from mercury poisoning and began to investigate the possibility. In June 1965, Hosokawa, who had retired from working as the director of Minamata Hospital in Kumamoto Prefecture, announced that the disease of Niigata patients was definitely Minamata disease.[2] The Japanese Ministry of Health and Welfare officially requested Niigata Medical University to identify the cause of the disease. The incident unfolded just like in Minamata. The initial report of Niigata Medical University, which mentioned the mercury waste of Showa Denko, thought to be the cause of the disease, was not published due to the stubborn obstruction of the Ministry of International Trade and Industry. In the mid-1966, the Ministry of Health and Welfare discontinued funding for Niigata Medical University. For the next three years, like in Minamata, patients in Niigata had to fight against huge barriers such as corporate responsibility evasion, government non-cooperation, and research manipulation by scholars who advocate corporate claims.

Finally, in April 1967, despite strong opposition from senior officials of the Ministry of Health and Welfare, the final report of the Niigata Medical University research team was published thanks to the efforts of some courageous members of parliament. However, Showa Denko insisted that even if the Ministry of Health

---

[1] https://en.wikipedia.org/wiki/Niigata_Minamata_disease.

[2] Island of Dreams: Environmental Crisis in Japan, Norrie Huddle and Michael Reich, Schenkman Books, 1987.

S. S. Park, *Eco-disasters*, https://doi.org/10.1007/978-981-96-8667-4_13

and Welfare accepted the report as an official government position, the company could not accept it. After the report of Niigata Medical University was published, the attitude of the victims changed radically. With the strong support of an organization called the Niigata Prefecture Democratic Union, the victims began legal proceedings against the company in June 1967. This was three months ahead of the lawsuit filed in the Yokkaichi case, and it was the first large-scale civil lawsuit against a polluter in post-war Japan. This lawsuit marked a significant turning point in Japan's environmental history. After that, the lawsuit of Yokkaichi asthma patients followed, and legal proceedings became a major means of resolving environmental pollution disputes.

At the end of 1967, patients in Niigata and Minamata began to correspond. A few months later, patients from Niigata visited Minamata to meet comrades in the same situation and gather strength. Their historic meeting inspired each other to fight. In January 1968, the two groups firmly resolved to hold out to the end in the fight against irresponsible environmental pollution-causing companies.

Eight months later, on September 26, 1968, the Japanese Ministry of Health and Welfare officially announced that the strange disease in Niigata was due to mercury poisoning, and the factory waste from Showa Denko, located 65 km upstream of the Agano River, was the cause. It was an official recognition by the government that Showa Denko provided the cause of the disease, just one year and three months after the victims in Niigata filed a lawsuit against Showa Denko. Coincidentally, Showa Denko was operating the same production facilities as the Minamata factory. Showa Denko discharged wastewater containing mercury into the river without treatment, and the disease appeared in people who continued to eat mercury-contaminated fish after going through the process of microbial methylmercury conversion, bioaccumulation, and biomagnification. The process of mercury emission, conversion, and movement was identical to that in Minamata.

The government's official recognition was not unrelated to the verdict in the Niigata trial on September 19, 1971. The court sided with the plaintiff's claims about the cause of the problem, the company's disregard for it, and the company's legal responsibility. Moreover, this verdict greatly contributed to the anti-mercury pollution movement gaining nationwide support. However, the victims in Niigata were dissatisfied with the court's decision to award much less compensation than they demanded. When their lawyers showed difficulty in resolving this legally, they decided to go directly to Showa Denko's Tokyo headquarters to protest. When Showa Denko's Chairman Haruo Suzuki refused to meet, they declared they would wait until he agreed to meet and began a sit-in at the company's office. Late that night, Chairman Suzuki changed his mind, met with the victims, and knelt before them.

Niigata's struggle process and achievement became an important reference to the Minamata verdict a year and a half later. The Niigata incident was a decisive factor in the Japanese government's recognition that the cause of the Minamata incident was mercury contained in factory wastewater. It took 15 years of reports of strange diseases among Minamata residents before they acknowledged the cause and knelt down after seeing a similar case. In Minamata, sea fish, and in Niigata, freshwater fish became the route for mercury to enter the human body. At that time, there were villages using river water as a water source in the middle of the Agano River, but no symptoms of mercury poisoning appeared here, and it was in the downstream where a lot of fish and shellfish were eaten. It only appeared in one place. This was a case that showed the phenomenon of bioaccumulation directly affecting mercury poisoning.

The Niigata incident was the first environmental lawsuit in Japan (June 1967), leading to the beginning of environmental disaster court battles, followed by Yokkaichi in the same year in September, Itai-itai in 1968, and Minamata in 1969. These incidents, which sacrificed numerous lives and became known as the four major environmental disasters, played a decisive role in the establishment of the Japan Environmental Agency in 1971. The total number of victims of the Niigata incident that have been identified so far was 2,665, most of whom lived their entire lives with incurable diseases and have now mostly passed away.

Chapter 14
Michigan PBB

The Michigan Chemical Company, located in St. Louis, Michigan, has manufactured a substance called PBB (PolyBrominated Biphenyls) since 1971 and sold it exclusively throughout the United States under the product name 'Firemaster'. PBB is a heat-resistant chemical used in telephones, electric typewriters, calculators, televisions, cars, etc., and is also used as a fire retardant when added to plastics. Before this product was developed, about 12,000 people died annually from fires in the United States, and about 300,000 people were severely burned. The federal government, believing that adding fire retardants to household appliances could reduce casualties from fires, legally mandated the use of PBB. Indeed, fire damage in the United States significantly decreased after the use of PBB. However, PBB, a useful chemical, caused a tremendous environmental disaster due to a few people's mistakes.[1]

At that time, the Michigan Chemical Company also manufactured and sold several other chemicals, and most of their product names ended with the word 'Master'. Among them, they manufactured a feed additive that promotes milk production in cows and sold it under the product name 'Nutrimaster'. One summer day in 1973, due to similar product names, the person in charge mistakenly put PBB in the feed additive bag and delivered it to a farm in central Michigan. PBB, which had been swapped with the livestock feed additive, was sold on the market. PBB is a deadly toxic substance that can cause cancer and neurological abnormalities when it enters the body of an animal or human. People ended up consuming milk and dairy products produced by cows raised with this PBB, and various damage symptoms appeared on the farm, but no one knew the cause until the end of 1975.

Eventually, the cause of this incident was revealed through the efforts of a farmer named Fredric Halbert. Halbert found it strange that milk production had decreased by 40% and feed consumption by 50% on his farm within 20 days in 1973, so he stopped using the suspicious feed additive. After stopping the use of the additive, feed consumption returned to normal, but milk production remained at 40% of the

[1] https://www.michigan.gov/mdhhs/safety-injury-prev/environmental-health/topics/dehbio/pbbs/history.

S. S. Park, *Eco-disasters*, https://doi.org/10.1007/978-981-96-8667-4_14

previous level. The cows' weight continued to decrease, and they showed symptoms of hair loss, infertility, and strange skin diseases. After six months, 24 out of 400 cows died, Halbert attempted to determine the cause by performing autopsies on the dead dairy cows. As a result of the autopsy, he discovered diseases in the kidneys and liver of the cows and suspected that the cause of the disease was in the feed additives. Eventually, he revealed that PBB, which had a fatal impact on livestock, was the cause after analyzing the additives that had been used.

There were as many as 87 ranches in Michigan that used PBB as a feed additive, and most of the dead dairy cows were sold to feed production factories. The feed produced here was used in pig and poultry farms, and the PBB residue in the feed caused a second chain of damage. Eventually, 30,000 dairy cows, 1.5 million chickens, 6,000 pigs, and 13,000 sheep died or were discarded. A large amount of eggs, cheese, butter, and milk produced during this period also had to be thrown away. Moreover, people who ate these showed symptoms of paralysis.

After this incident, on November 4, 1974, the U.S. Food and Drug Administration (FDA) stipulated the permissible level of contamination, which was 0.3 ppm for milk and meat and 0.05 ppm for eggs and feed. However, according to the announcement of the Michigan Department of Agriculture on October 7, 1974, the PBB concentration in milk due to this incident was 595 ppm, eggs were 59.7 ppm, feed was up to 13,500 ppm, the PBB concentration in chicken bodies was 4,600 ppm, and the PBB concentration in calf bodies was 2,700 ppm.

A more serious problem occurred in the process of disposing of the dead livestock and livestock products. Even at that time, knowledge about environmental pollution was lacking, and accident handling measures were not systematically carried out. People, unaware that PBB does not easily decompose, buried dead livestock and livestock products anywhere, contaminating nearby soil, groundwater, and rivers. Efforts were made later to purify the contaminated soil, groundwater, and rivers, but PBB continued to be detected for a long time. The direct loss from this accident was estimated to be between 75 and 100 million dollars, but considering the cost of disposing of dead livestock and the cost incurred by environmental pollution, the economic damage was much greater. The Michigan Department of Agriculture filed a lawsuit against the chemical company for 270 million dollars and eventually settled for 20 million dollars. In addition, individual lawsuits were filed from farms, but the physical and mental damage received by the ranch owners and their families could never be converted into money.

An incident of toxic substances entering livestock feed also occurred in Belgium in January 1999. Here, an accident occurred where materials containing toxic substances such as PCBs and Furan, which were included in the raw materials for livestock feed, were mixed with substances in transformers. Over 1,700 Belgian livestock farmers purchased feed containing toxic substances and fed it to chickens, pigs, and cows. Initially, there was no reaction from the livestock, but in March, high concentrations of PCBs and Furan began to be detected in chicken meat and eggs, revealing the truth of the incident. By the end of May, Belgian authorities had disposed of all products containing eggs, including all poultry, mayonnaise, cakes, and biscuits. All suspicious livestock were also slaughtered, and measures were taken to completely ban the import of Belgian meat products in countries around the world.

In 1969, in Alamogordo, New Mexico, USA, there was an incident where people who ate pork raised on feed contaminated with pesticides went blind. A farm in Alamogordo raised pigs with grain scraps obtained from a grain warehouse in the nearby town of Texaco, and one day, 14 pigs began to go blind, and eventually 12 died. The farm, which continued to raise pigs without knowing the cause, also slaughtered and ate the grown pigs. A few months later, mysterious symptoms began to appear in the family of the house. An 8-year-old daughter fell off the school bar and injured her back because she couldn't balance herself, but the local hospital couldn't determine the cause of her imbalance. While the child was hospitalized, two other children also showed similar symptoms and were taken to a nearby metropolitan hospital. An investigation of the cause again concluded that there was a connection between the pigs going blind and the children's symptoms. Soon after, mercury was detected in the grain scraps fed to the pigs, and it was revealed that this was due to the pesticide 'Panogen', a seed disinfectant containing methylmercury, sprayed in the grain storage warehouse. After that, the children went blind forever.

Panogen is a pesticide used extensively in farms to disinfect grains to prevent mold. The mercury contained in it not only causes physical abnormalities when poisoned, but also is a harmful heavy metal that causes abnormalities in the nervous system and behavior. What's more surprising is that after this incident was reported through the media, it was revealed that similar incidents had occurred in various places in the United States, but the cause was not identified and was buried. When damage cases were reported one after another from various parts of the United States, the Department of Agriculture stopped selling mercury-based pesticides. However, the manufacturer sued the Department of Agriculture, arguing that mercury-based pesticides were essential for increasing food production. The manufacturers' claim that there is no direct evidence that these incidents originated from the use of mercury-based pesticides was accepted, and the sale of mercury-based pesticides was eventually allowed again.

The Michigan PBB incident greatly contributed to the institutionalization of more thorough management of the production, transportation, and disposal of toxic chemicals in the United States. In 1976, the United States enacted the Resource Conservation and Recovery Act (RCRA) for the management of hazardous substances, introducing the 'Manifest System' that requires producers to verify and report the entire process from transportation and use to disposal of hazardous substances, and to take responsibility in case of disaster. In other words, the system obliges producers to take responsibility for toxic chemicals 'from cradle to grave (from production to disposal)', and this system is adopted in most advanced countries around the world.

Chapter 15
Kanemi Oil PCB

In March 1968, residents of a fishing village called Damanoura on Fukue Island, 120 km west of Kyushu, Japan, began to experience mysterious symptoms.[1] They lost their appetite and were constantly drenched in sweat. Moreover, they were so fatigued that it was difficult to work, and grotesque boils broke out on their bodies, from which a foul-smelling pus flowed. Their vision deteriorated, and mucus flowed from their eyes, causing their eyelids to stick together. The children could not go to school, and the entire family was unable to work, leaving their livelihood in jeopardy. It was as if a divine curse had descended on the village.

The biggest problem was the boils that covered their bodies. As June passed, popping the boils became the most important task of the day, and eventually, they had to spend all day doing it. The pain was unbearable. Doctors could neither identify the cause of this strange disease nor treat it. Since this fishing village was far from pollution, it was suspected that they had contracted a terrible disease like leprosy or syphilis. People were caught in despair as they argued over the cause of the disease, but in the end, they could not find any answers.

On the evening of October 10 that year, news about the strange disease that broke out in a village in northern Kyushu was first reported in Damanoura. The cause of the disease was said to be contaminated cooking oil. Surprisingly, the symptoms were exactly the same as those of the villagers in this fishing village. The person who identified the cause of the disease was a Kyushu Electric Power Company employee named Datashiro Kunitake. When he and his family contracted an unknown disease, he persistently searched for the cause. He happened to meet patients with the same symptoms as him at a dermatology clinic and found out their common point. They all liked fried food and used the same brand of cooking oil. So, Datashiro began to suspect the Kanemi cooking oil he used at home. This rumor spread widely, and when Asahi newspaper reporter Abe Masaki wrote an article about it, the cause was revealed. As a result of the investigation, it was confirmed that the disease broke out

[1] Island of Dreams: Environmental Crisis in Japan, Norrie Huddle and Michael Reich, Schenkman Books, 1987.

S. S. Park, *Eco-disasters*, https://doi.org/10.1007/978-981-96-8667-4_15

in people who ate food cooked with cooking oil manufactured and sold by Kanemi Company, and that the Kanemi cooking oil contained a harmful substance called polychlorinated biphenyl (PCB).

PCB is the name of a group of chemical substances, and its commercial name is Arochlor, Phenochlor, also known as Chlopher, is a substance that does not decompose even when heat is applied and has excellent electrical insulation, so it was widely used in electrical products. However, when PCB is released into the environment, it does not decompose and remains intact, accumulating in the bodies of organisms and becoming toxic. When PCB enters the body, it can cause not only skin diseases but also liver diseases, endocrine system abnormalities, neurological disorders, and in severe cases, death. PCB has a molecular structure similar to DDT, so the toxic phenomena are very similar. However, in terms of the intensity of toxicity, PCB is known to be much more dangerous than DDT.

At that time, the Kanemi company used PCB as a heating medium in the manufacture of edible oil, and the cause of this incident was that the heating pipe corroded and PCB flowed into the edible oil.[2] Edible oil was a staple food for the Japanese who loved fried food, and as a result, many people were poisoned by the PCB in Kanemi's edible oil. The investigation revealed that the company had poor safety management and equipment facilities for PCB at the time. Some victims claimed that the company knew all these facts in advance, citing that the edible oil was sold at a drastically cheap price, mainly on remote islands.

As a result of this incident, about 14,000 people complained of physical abnormalities such as memory loss, hair loss, decreased libido, fatigue, pharyngitis, headache, abdominal pain, dizziness, deformities of bones and muscles, jaundice, lethargy, erectile dysfunction, and discoloration of skin, nails, mouth, and around the eyes, and among them, 1,068 people had severe symptoms that required special treatment. Also, about 90% of the mothers who ingested PCB gave birth to children with dark or grayish skin, which is because, like mercury, PCB penetrated the placenta and concentrated in the fetus, causing the skin to darken and growth disorders. Also, about 1 million chickens were poisoned by eating feed containing Kanemi oil, and about 700,000 of them died. To this day, no effective treatment for PCB poisoning has been developed. Peripheral symptoms such as headaches and abdominal pain can be alleviated with medication, but there is still no way to neutralize or remove the accumulated PCB in the body. Eventually, this substance remains in the human body and the damage will continue for a lifetime.

The Kanemi oil incident started with food contamination, but it became a good example of the toxicity of PCB as it was highlighted as a harmful environmental pollutant in various places around the world, including the Great Lakes and Hudson River in the United States. In response to this incident, Japan began to regulate substances that do not decompose in the environment and are harmful to the human body, such as PCB, as specific chemicals by law in 1972. In accordance with this law, special management was carried out when manufacturing or importing these chemicals.

---

[2] https://en.wikipedia.org/wiki/Yusho_isease.

In 2001, the United Nations adopted the Stockholm Convention to regulate the movement of non-degradable organic pollutants (POPs: persistent organic pollutants) such as PCBs between countries, and it began to take effect as an international agreement in 2004. This is to prevent non-degradable pollutants from being discharged to the natural world because they do not decompose and travel around the earth to continuously damage the human body and ecosystem.

Chapter 16
Woburn Groundwater Contamination

Woburn, Massachusetts, in the eastern United States, was one of the places where industrialization first began in the 19th century. The leather industry flourished here in the early days, leading to the establishment of many factories. The development of the leather industry led to the spread of other industries, and later various chemical factories were established here. However, the leather industry began to decline in the 1960s, and one by one, they closed their doors, and instead, large corporate chemical factories came in and replaced the existing industries.

In 1972, a 3-year-old child named Jimmy living here was diagnosed with leukemia. Jimmy's mother, Anne Anderson, believed that her son had fallen ill due to contaminated drinking water. At that time, leukemia was a rare disease that barely affected 4 out of 100,000 children, but there were unusually high number of leukemia patients in Woburn. Therefore, Anne publicly claimed that the groundwater used here was clearly related to leukemia, but people around her only considered her a fussy and unusual mother.

Long before Jimmy was diagnosed with leukemia, people here often complained about drinking water. The Aberjona River, which flows through this place, has been floating with various pollutants for over 150 years, and the groundwater used for drinking water often caused problems with unusual tastes and smells. Especially after two wells were built in 1964 and 1967, the taste of the water noticeably worsened. The safety of these two wells continued to be controversial, and they were repeatedly closed and operated. Despite the opposition of the residents, the wells were closed, but when a drought hit or water demand increased, they had no choice but to operate again.

Then, in the spring of 1979, the Woburn Police Department discovered 184 industrial waste drums abandoned in a vacant lot in the northern region. The scene of the incident was just about 500 m away from the two wells. When the water quality of the two wells was tested, it was found to be heavily contaminated with the industrial solvent trichloroethylene (TCE).

S. S. Park, *Eco-disasters*, https://doi.org/10.1007/978-981-96-8667-4_16

In addition, an epidemiological survey of the residents of this area confirmed that 28 cases of leukemia had occurred since 1964. This figure was seven times higher than other areas, considering the size of the town. It was also found that most of the patients lived near the well in question, and all of them drank contaminated well water as drinking water. The authorities permanently shut down the well, which was contaminated with TCE and Perchloroethylene (PCE).

A few days after Christmas in 1981, Jimmy eventually died at the age of 12, and the child's mother filed a lawsuit against three companies considered to be the parties that contaminated the drinking water along with seven other households in the same village.[1] The sued companies, UniFirst, Beatrice Foods, and Grace & Company, had been operating factories using various harmful chemicals including TCE here for a long time. Among these companies, UniFirst managed to get out by paying about a million dollars in settlement money before the lawsuit began, and the remaining two companies had to continue the truth game about whether they were responsible for well pollution and whether the pollutants caused leukemia.

During the lawsuit, two professors from the Harvard School of Public Health published a three-year investigation report on leukemia that occurred in Woburn, concluding that the two wells in this place were contaminated with various harmful substances and that the well water was related to leukemia. In response, the defendants mobilized several professors and doctors to argue that TCE is unrelated to leukemia and skeptical about its carcinogenic potential. However, scholars supporting the plaintiff judged that even a small amount of TCE could cause abnormalities in the immune system if exposed for a long time, conducted blood tests on residents, and sought data to support their claims. As a result, carcinogens were found in the blood of Woburn residents, and an opinion was submitted to the court that TCE detected from Woburn's well caused immune dysfunction and leukemia, among other serious diseases.

The next step was to find out who polluted the groundwater, whose order it was, and how much it was polluted. The plaintiff's lawyer persistently investigated the employees of the sued companies and revealed that these companies had illegally disposed of harmful chemical wastes including TCE in vacant lots and pits. They also suggested the possibility that the things thrown away here flowed to the Aberjona River and contaminated the riverside wetlands with toxic substances. Even if they couldn't directly conclude that the TCE discarded by these companies caused leukemia, it was becoming increasingly clear that the contaminated drinking water exacerbated leukemia.

The fact that the plaintiff had to further prove was how the harmful substances that flowed into the river contaminated Woburn's groundwater. In front of the jury, the lawyer for Grace and Company admitted that the company's employees had dumped TCE on the ground, but emphasized that the amount was so small that it would not have affected the well, and even if it had entered the well, it would not have been enough to pose a risk to the human body. However, the lawyer for Beatrice Foods denied all charges, stating that the company had never used TCE and

[1] https://en.wikipedia.org/wiki/Anderson_v._Cryovac,_Inc.

had never illegally dumped hazardous waste. The plaintiff's lawyer sought to prove, through expert consultation, that the defendant's factory land was contaminated with hazardous chemicals and that this had flowed into the groundwater, contaminating Woburn's well by the late 1960s. However, the scholars, unfamiliar with the courtroom atmosphere, were inexperienced in properly conveying their opinions against the attacks of the seasoned opposing lawyers. The defendant pushed the investigation results so far as false and disputed the qualifications of the witness.

The final witness of the trial was George Pinder, a groundwater hydrologist from Princeton University, who had to prove that harmful substances such as TCE discharged from the factory had entered Woburn's well before leukemia or other related diseases occurred. His testimony went smoothly, and his opinion that the two companies were responsible for well contamination was adopted as evidence in court. However, there was a mistake in presenting the time it took for the pollutants to reach the well, and the opposing lawyer raised a counterargument. To make matters worse, Professor Pinder reversed his claim in his testimony about where the well water came from, and the trial turned against the plaintiff.

The Aberjona River was located between the Beatrice Foods factory area and the well, and the water level of this river dropped significantly after the well began to draw groundwater. Based on this, the defendant argued that the polluted substances from the Beatrice Foods factory did not enter the well because the river water entered the well independently of the groundwater. In response, Professor Pinder argued that the well water came from the groundwater, not the river, and that the river's water level dropped because the contaminated groundwater was pumped up from the well and then the river water entered the groundwater. This was theoretically valid, but it was not cited in the trial because there was no scientific evidence.

After a tumultuous legal battle spanning 78 days, the jury decided in June 1986 to dismiss the claims against Beatrice Foods and hold Grace and Company solely responsible for well contamination. Grace and Company agreed to a compensation of $8 million, which was far less than the amount expected when the lawsuit was first filed. Later, the U.S. Environmental Protection Agency (EPA) conducted a pump test on the well and, surprisingly, concluded that both Grace and Company and Beatrice Foods had contaminated Woburn's groundwater. In particular, the factory site of Beatrice Foods was identified as the major source of pollution. The jurors and judge who participated in the trial made a wrong judgment by ignoring the expert's testimony and not acknowledging the fact that the groundwater had contaminated the well on the other side of the river. The plaintiff's lawyer soon appealed, discovering that the lawyers and witnesses for Beatrice Foods had perjured themselves and tried to conceal evidence, and tried to hold them accountable. However, the plaintiff lost again due to insufficient evidence.

After this incident, the U.S. Environmental Protection Agency announced a plan to purify Woburn's groundwater. The EPA sued Grace and Company and Beatrice Foods to recover the purification costs, and these companies eventually agreed to share the costs. The EPA has implemented the largest purification plan in the history of the northeastern New England region, costing a total of $69.4 million over 50 years.

This case stands as a significant record of individuals who fought a difficult legal battle against corporate negligence leading to severe environmental contamination and considerable harm to residents' lives. Unfortunately, the story of a small David who devoted his passion to protecting the truth and faith against a huge Goliath is more dramatic and touching than any other novel, although those who suffered severe physical and mental damage failed to receive due compensation. After that, Civil Action, meaning a civil suit dealing with the case, was published in the United States in 1995 and was a bestseller, and was produced as a film of the same name and introduced to countries around the world.[2] The main characters of the film, plaintiff's lawyer Jan Schlictmann, and the actor who plays Anne Anderson, who led the lawsuit, were John Travolta and Kathleen Quinn, respectively. Along with Erin Brockovich, which is based on the Hinckley Chrome case, the film is considered one of the representative films about the true story of an environmental lawsuit court.

[2] https://en.wikipedia.org/wiki/A_Civil_Action_(film).

Part IV
Sick and Poisoned Nature

Chapter 17
Clear Lake DDD

Clear Lake, located about 150 km north of San Francisco, California, USA, has long been popular with fishermen. However, this lake was extremely annoying to its visitors due to the explosive increase of small insects called gnats in the summer.[1] Therefore, the lake managers have been considering ways to eradicate the gnats that become a problem in the summer. Thus, various methods have been attempted since 1916, such as using fish to eat the gnat larvae or spraying insecticides that were beginning to be developed at the time. However, these efforts yielded little success.

In the midst of this, the lake managers massively sprayed a newly developed insecticide called Dichloro-Diphenyl-Dichloroethane (DDD) in 1949.[2] DDD is an organochlorine chemical with a molecular structure very similar to DDT, and when sprayed in nature, it does not easily decompose and remains for decades. The managers confirmed through preliminary experiments that DDD is relatively harmless to fish, but its insecticidal effect on gnat larvae is excellent. After that, gnats were hardly found in Clear Lake, but they began to increase tremendously again in 1951. In 1954, they massively sprayed DDD again, increasing the intensity by 50%. That winter, over 100 grebes, birds that feed on fish from the lake, were found dead. The dead birds were examined, but there was no sign of an epidemic. Grebes flock to this area in winter to lay their eggs. With their white necks and black heads held high, groups of grebes, swimming gracefully on the lake's surface with their chicks, used to create a spectacular winter scene at Clear Lake.

In 1955 and 1956, the gnats increased explosively again. Therefore, in 1957, DDD was massively sprayed at the same intensity as in 1954, and more grebes were found dead on the lakeshore. Again, there were no signs of an epidemic. However, an analysis of the grebes' adipose tissue revealed a high concentration of DDD, reaching 1,600 ppm. The maximum concentration of DDD introduced into the lake was 0.02 ppm. The concentration of DDD in the grebes' bodies was more than

[1] https://en.wikipedia.org/wiki/Clear_Lake_(California).

[2] Silent Spring, Rachel Carson, 1962.

S. S. Park, *Eco-disasters*, https://doi.org/10.1007/978-981-96-8667-4_17

80,000 times higher than in the water. To understand why, they investigated the DDD concentration in other organisms living in the lake, and found that plant plankton had 5 ppm, herbivorous fish had 40–300 ppm, and predatory fish that ate other fish, such as catfish, had It showed a high level of 2,500 ppm. Following the food chain, DDD in the water was concentrated in phytoplankton, phytoplankton in herbivorous fish, herbivorous fish in carnivorous fish, and then in grebes. Through this incident, scientists confirmed for the first time that pesticides are amplified along the food chain (Biomagnification). And for the 10 years that DDD was sprayed, fishermen caught fish concentrated with toxic substances in exchange for easily solving the annoying gnats, and cooked and ate them with friends or family. DDD is a substance that destroys the adrenal cortex, which secretes a hormone called cortin, causing endocrine abnormalities when it enters the human body.

DDD spraying in Clear Lake has been discontinued since 1959, but even more surprising facts have been revealed. Even 23 months after the spraying was discontinued, 5.3 ppm of DDD was detected in phytoplankton. For almost two years, plankton were born and disappeared continuously, but DDD remained. Of course, it was continuously detected in fish and grebes. The number of grebe nests, which was over 1,000 before DDD was applied to the lake, dropped dramatically to about 30 in 1960. Even in the remaining nests, the number of grebes decreased significantly because hatching was not done properly. It confirmed that once sprayed pollutants remain in nature and circulate persistently.

The Clear Lake incident was a decisive factor in the birth of "Silent Spring", the greatest environmental masterpiece of the 20th century. The author, Rachel Carson, had already published "Under the Sea Wind" (1941), "The Sea Around Us" (1951), "The Edge of the Sea" (1955), and "Our Ever-Changing Shore" (1957), conveying the beauty of nature and the wonder of life to many readers. Her view of nature consistently shown in all these books is that humans are only a part of nature, and the difference between humans and animals is the ability to change nature.

The Clear Lake incident was shocking to Rachel Carson and was considered an appropriate example to explain her view of nature. She also learned for the first time through this incident that even low concentrations of pesticides sprayed on nature are concentrated at high levels in birds, the last point of the ecosystem food chain, leading to death. She wanted to inform people of this fact, which was quite shocking at the time. She began writing in the spring of 1958 and published "Silent Spring" in 1962. Through this book, she denounced the tragedy of birds disappearing due to the overuse of pesticides, and the arrival of spring without the sound of birds. "Silent Spring" caused a huge stir in American society as soon as it was published. The book awakened many people to the consequences of excessive pesticide use. From then on, debates about the benefits and harms of pesticides began, and scientific investigations were conducted. President Kennedy at the time formed a scientific advisory committee to review the issues raised in the book. However, it was not easy for the government to impose a ban on the use of pesticides due to the benefits of increasing food production and pest control.

The most widely used pesticide at the time was DDT, which brought about a revolution in agriculture when it was first introduced in the early 1940s. DDT was first synthesized in 1874 by Othmar Zeidler of Austria and its toxicity to pests was revealed in 1939 by Paul Müller of Switzerland. From then on, this pesticide began to be used on a large scale in all agricultural fields. The initial effects of DDT were tremendous. It greatly increased agricultural production by protecting crops from pests, saved millions of lives from malaria and yellow fever by eradicating mosquitoes, and significantly reduced the incidence of typhus. Before the use of DDT, about 200 million people worldwide were threatened by malaria each year, and about 2 million people lost their lives to the disease. People believed that DDT would soon eradicate all pests, and Swiss scientist Müller, who clarified the efficacy of DDT, even won the Nobel Prize in Chemistry in 1948.

In the 1940s and 1950s, DDT was used without limit in almost all countries, including the United States, for pest control. During World War II, DDT was an essential military supply for the U.S. military. In 1943, the U.S. Army greatly reduced typhus outbreaks in Italy by spraying a large amount of DDT.

However, starting with the Clear Lake incident, many cases of environmental damage caused by non-degradable pesticides began to be reported. Pesticides do not decompose easily, so once sprayed into nature, they remain in the soil for decades and are concentrated in living things. In North America, for example, after the use of pesticides, the number of many wild animals and birds has declined sharply, and fish have become inedible to humans or die. In addition, pesticides concentrated in the human body not only caused birth of deformed babies, but also caused various diseases, including cancer. In particular, as the Bald Eagle, the symbol of the United States, was endangered, Americans had to realize the environmental damage caused by pesticides.

With the publication of "Silent Spring", regulations were imposed on the use and purpose of pesticides, and the annual production of DDT significantly decreased. In addition, the development of pesticides that can easily decompose in nature and effectively control pests began. In 1972, the U.S. Environmental Protection Agency finally banned the use of DDT and Aldrin nationwide, and in 1974, Dieldrin, and in 1975, Chlordane and Heptachlor were also banned. In most developed countries, the use of all persistent organochlorine pesticides, including DDT, is currently banned, and relatively easily decomposable organophosphorus pesticides are being used. However, DDT is still being used in developing countries. This is because if the use of DDT is discontinued, millions of people, especially children, are easily exposed to the risk of diseases such as malaria and starvation.

The Clear Lake incident had a huge impact on human society when it gave birth to Silent Spring. As a result, the development and use of all synthetic chemicals revolutionized, and international efforts to prevent the extinction of living things began. It was also an opportunity to look back on the development of science and technology that was recklessly going forward at the time. When developing a new chemical substance, mankind introduced a system that allows its use after confirming its safety by reviewing in advance the impact of that substance on the natural ecosystem. This system, called 'Ecological Risk Assessment', started in the United States in the 1980s and is currently being implemented in most developed countries. Looking back, the death of grebes flock in Clear Lake in the 1950s may have prevented the 'Silent Spring' that might arrive on Earth.

Chapter 18
Vietnam War Herbicide

The Vietnam War, spanning the 1960s and 1970s, served as a catalyst for unifying a divided nation, but on the other hand, it was a war that resulted in numerous human casualties and material damage. Among them, environmental damage caused by defoliants is still being reported even half a century after the war. During the Vietnam War, a terrible nuisance to the United States military was the counterattacks from enemies hidden in the jungle. Also, the jungle was a major obstacle to the American style of warfare based on large weapons. The United States, which had experience fighting in the jungle during the South Pacific battles of World War II, began using defoliants to remove the jungle shortly after its involvement in the Vietnam War in 1962.

The United States is recorded as having developed and used chemical substances for jungle removal in the New Guinea battle in 1944 and 1945. In the early stages of the Vietnam War, the United States used defoliants known as Agent Green, Pink, and Purple on a small scale. However, following the Gulf of Tonkin Incident in 1964, the United States became fully involved in the Vietnam War and from 1965, developed and massively sprayed a highly effective defoliant known as Agent Orange.[1]

Agent Orange was named after the orange-colored barrels that contained the substance. The defoliant, made by mixing two types of chemicals, 2,4,5-T and 2,4-D, in a 50:50 ratio, was very effective in removing the jungles of Vietnam. It was also called a defoliant because the leaves dried up and died when sprayed on the jungle.

The U.S. Air Force assigned the task of spraying defoliants by plane all over Vietnam to a special operation unit called Ranch Hand. In the Vietnam War, where there was no distinction between front and rear due to the emergence of Viet Cong (guerrilla army) hidden in the jungle, the entire jungle area of the country was targeted for defoliant spraying, and it was also intensively sprayed on farmland, the enemy's source of food supply. Agent Orange was very effective in killing beans, peanuts, potatoes, mangoes, etc., the staple food of the Vietnamese. However, when

[1] https://en.wikipedia.org/wiki/Agent_Orange.

S. S. Park, *Eco-disasters*, https://doi.org/10.1007/978-981-96-8667-4_18

Agent Orange had little effect on killing rice, the United States developed a defoliant containing highly toxic arsenic called Agent Blue and sprayed it on rice paddies.[2]

It is recorded that the United States sprayed 72 million liters of defoliants over 10 years in the Vietnam War. 90% of the spray was used to remove jungles, 8% was sprayed on enemy farmland to starve their food supply, and the remaining 2% was sprayed around our military positions to guard against enemy invasions. The U.S. military achieved great success in jungle warfare through this defoliant spray, but they did not anticipate the enormous environmental disaster that would follow. 2,4,5-T, which makes up half of the defoliant, produces a highly toxic chemical called dioxin as a byproduct during its manufacturing process. Due to this dioxin, Vietnamese and participating soldiers suffered serious damage after the war. At the time of spraying, the harmfulness of dioxin was not scientifically established, and the proportion of impurities produced in the 2,4,5-T manufacturing process was not known, so there was no effort to remove this harmful impurity from the herbicide.

The question of whether it was safe to spray defoliants without knowing the harmful effects of dioxin was first raised in 1966 by the National Academy of Sciences in the United States. The Academy warned that if herbicides continued to be sprayed in large quantities in the Vietnam War, it would cause enormous damage to humans and the natural ecosystem. In 1967, it proposed an investigation into the impact that would eventually appear in Vietnam due to defoliant spraying to the U.S. Department of Defense. That same year, over 5,000 American scientists submitted a petition to the Presidential Science Advisory Committee to ban the use of defoliants. They strongly condemned the risks of using chemical substances in war and the inhumane act of starving the enemy's food supply with herbicides.

As the warnings continued, the U.S. government was afraid that the defoliant spraying for jungle removal would be mistaken for the use of chemical weapons and receive international criticism. To avoid this misunderstanding, the military authorities used the term Herbicide Orange, which sounds less military, instead of Agent Orange. Meanwhile, the U.S. Department of Defense contracted with the Midwest Research Institute to investigate the problems with the use of defoliants. However, without an on-site survey in Vietnam, the institute reviewed over 1,500 published papers on herbicides and interviewed over 140 experts, and announced the results after just three months. The conclusion drawn by the Midwest Research Institute was that there was no long-term damage from the use of defoliants. It was also stated that there were no existing research findings on the effects of the arsenic contained in Agent Blue, which was used to kill rice. Later, the institute was criticized for not even attempting to investigate the long-term damage caused by defoliants.

In 1968, the then U.S. Ambassador to South Vietnam, Ellsworth Bunker, invited Dr. Fred Tschirley from the U.S. Agricultural Research Service to Vietnam to investigate the damage caused by the use of Agent Orange. Dr. Tschirley reported that the use of Agent Orange was causing serious ecological damage, particularly destroying the jungles of coastal areas, which are major habitats for fish and birds. In 1969, Professor E.W. Pfeiffer of the University of Montana and Professor G. H. Orians of

[2] How to Survive in America the Poisoned, Lewis Regenstein, Acropolis Books, 1986.

the University of Washington visited to Vietnam to investigate the ecological damage caused by the use of Agent Orange. These two professors also confirmed the findings of Dr. Tschirley and warned that the use of Agent Orange should be ceased.

The same year, decisive research results on the use of Agent Orange came out in the United States. It was announced that animal experiments showed that 2,4,5-T could cause deformities. Based on these experimental results, the U.S. Department of Agriculture (which was responsible for environmental damage caused by pesticides and herbicides before the establishment of the U.S. Environmental Protection Agency in 1970) banned the use of 2,4,5-T on farms near residential areas in the U.S. In October 1969, President Nixon Presidential Scientific Advisory Committee warned that defoliants should be sprayed only in uninhabited places in Vietnam. The U.S. State Department dismissed this warning, arguing that such a warning was meaningless because Agent Orange had never been sprayed where people live. However, five months later, in March 1970, the State Department expressed its intention to reduce the use of Agent Orange in Vietnam by 25%, and a month later, it announced a total ban on the use of Agent Orange in Vietnam.

In April 1972, the U.S. Air Force disposed of the remaining 1.37 million gallons (5.106 million liters) of Agent Orange. The U.S. Air Force, which had no suitable place to dispose of the herbicide containing dioxin, chose to install an incinerator on a ship and go out to sea to incinerate it at high temperatures. Dioxin decomposes into harmless substances when burned at temperatures above 2,000 °C. This method was possible because it was before the London Convention, which was adopted the same year, came into effect. The London Convention, which came into effect in 1975, was created under the leadership of the International Maritime Organization (IMO) to prohibit marine dumping and incineration of waste, and the U.S. also ratified it.

In reality, the period during which herbicides containing dioxin were sprayed in Vietnam lasted only five years. However, due to this dioxin, the stillbirth rate in Vietnam surged after the war, and the incidence of birth defects was ten times higher than before the war. Additionally, the effects on soldiers who participated in the war continue to this day. These veterans have reported higher cancer rates compared to the general population, an increased occurrence of birth defects in their children, and a high prevalence of mental health issues.

Among the Korean soldiers who participated in the Vietnam War, 126,479 people complained of symptoms of Agent Orange damage, and 86,674 of them were designated as sequelae patients or similar sequelae patients, receiving pensions and allowances from the government. In the United States, about 240,000 people, and in Australia, about 1,200 people are registered as victims of Agent Orange in the Vietnam War, receiving compensation from the government. Most of them exhibit mental health issues along with physical ailments, and, as with diseases caused by other forms of pollution, healing is impossible.

The Vietnam War left about 20 million bomb craters as traces on the territory. Vietnam lost 40% of its farmland to Agent Orange spraying and bomb craters, and lost half of its jungle. This vividly illustrates the horrific environmental destruction and pollution caused by war. The International Union for Conservation of Nature (IUCN) warned that if the post-war natural environment of Vietnam is left as it is, it will become an irrecoverable wasteland, and estimated that to prevent this, 500 million trees should be planted across Vietnam. The damage from Agent Orange reminded us of the lesson that we must thoroughly investigate what harmful substances are included in the chemicals sprayed on nature in advance.

Chapter 19
Amoco Cadiz Oil Spill

The beautiful sandy beaches of Brittany, France, are world-renowned tourist destinations, attracting numerous Europeans during the summer. On the night of March 16, 1978, a 220,000-ton oil tanker, the Amoco Cadiz, deviated from its course and hit a reef in the waters near Portsall, Brittany. The ship, owned by the American oil company Amoco, was carrying 1.6 million barrels of crude oil from the Middle East to Europe.[1]

The stranded oil tanker began to sink, slowly spewing crude oil into the sea. The weather was fine, but the accident occurred due to a combination of a malfunction in the navigation key and the captain's mistake. However, if immediate rescue measures had been taken after the accident, it would not have led to such enormous damage. However, much time was wasted in negotiations to hire a tugboat to salvage the sinking oil tanker, and the rescue ultimately failed. Not only that, but measures to prevent the spread of the leaked oil were not promptly implemented. The Amoco Cadiz sank deep into the sea, and all the oil on board was leaked.

As a result of this accident, 15,000 square kilometers of sea were covered with a thick oil slick, and 320 km of beautiful beaches turned into stinking lumps of oil. Just a few days after the accident, millions of seagulls and sea ducks died, their bodies covered in oil, and all sorts of coastal creatures, such as clams, crayfish, and sea urchins, died en masse. In addition, 70% of the seaweed that lived along the coast was covered in oil and turned into garbage, and the production of oysters in the area decreased by 80%.

At the time, France mobilized soldiers and young volunteers to collect the thick oil in order to save the dying coast. However, it took a long time to restore the area to its original state, and enormous damage had to be endured. The damage caused by this accident amounted to a total of \$390 million, including \$46 million in losses to the coastal fishing industry, \$192 million in lost tourism revenue, and \$142 million

[1] https://en.wikipedia.org/wiki/Amoco_Cadiz_oil_spill.

S. S. Park, *Eco-disasters*, https://doi.org/10.1007/978-981-96-8667-4_19

in cleanup costs. This accident is recorded as one of the largest oil spills of the 20th century, causing enormous damage.

Oil spills from tankers are one of the serious marine environmental problems facing modern humanity. There have been several large-scale accidents of a similar scale to the Amoco Cadiz incident to date. It has happened before, and it will continue to happen. On March 24, 1989, in the Prince William Sound of Alaska, USA, the Exxon Valdez incident resulted in the spillage of about 38,000 tons of crude oil, causing serious damage to the coastal ecosystem and fisheries. The cost of cleaning up the spilled oil was approximately $843 million. Then, on January 5, 1993, the Braer, a tanker carrying 92,000 tons of crude oil, lost its course and collided with a reef off the coast of the Shetland Islands in the North Sea while heading towards Canada, resulting in the total spillage of its oil cargo.

Oil spills are not just caused by tankers. On June 3, 1979, a pipe broke at an oil field owned by Pemex, the Mexican National Oil Company, located in the Gulf of Mexico, causing 440,000 tons of crude oil to spill into the sea. It took a whopping nine months to stop the oil from overflowing from the continental shelf into the sea. This accident caused massive damage to the rich fishery resources and marine ecosystem of the Gulf of Mexico. On April 20, 2010, another major oil spill occurred in the Gulf of Mexico. A massive explosion occurred at the Deepwater Horizon drilling facility. The accident caused a fire, killing 11 of the 126 onsite personnel and injuring 18. It is also estimated that a hole in the drilling pipe caused about 602,000 tons of crude oil to spill.

Today, more than 1,000 large and small oil spills occur every year around the world, and about 100 million tons of crude oil flows into the sea during oil tanker accidents or ship transportation. It is estimated that between 3 and 6 million tons of oil flows out every year, when the outflow from land and oil storage tanks or oil fields are combined. Some of the oil that enters the ocean evaporates into the atmosphere or decomposes in the water, but a large amount is deposited under the sea or on the coast and remains for a long period of time, and it takes decades for the damaged ecosystem to recover.

Oil is an indispensable resource for our lives and industries. However, it is also a major pollutant of all environmental media, including water, air, and soil. Among all forms of pollution, oil spills from tankers in sea, in particular, result in the most severe and direct environmental damage. Swift response can significantly mitigate the impact of accidents if possible. Unfortunately, most incidents occur in oceans where adequate oil spill response systems are not in place, leading to massive environmental disasters.

Chapter 20
Love Canal

Niagara Falls, located on the border between the United States and Canada, is widely known for its grandeur as the largest waterfall in the world. Connecting Lake Erie and Lake Ontario, two of North America's Great Lakes, this waterfall features a significant drop due to the 85-meter elevation difference between the lakes. If this waterfall did not exist, it would be possible to navigate ships freely from the interior of the United States to the Atlantic coast.

In 1892, an ambitious entrepreneur named William T. Love capitalized on this potential and initiated a plan to construct a 10-kilometer canal in the area to facilitate ship navigation and establish a hydroelectric power plant. At that time, direct current (DC) electricity wasted a significant amount of power during long-distance transmission. Love envisioned that by building a power plant here to provide cheap and abundant electricity, he could attract many factories and foresee a large city with a population ranging from 200,000 to 1,000,000. Electricity was considered the highest value in civilization at the time, and Thomas Edison had already succeeded in the practical application of hydroelectric power in 1882 in Appleton, Wisconsin. Consequently, Love's canal construction plan received approval and enthusiastic support from the state government.

However, when the canal was about 1.6 km long, an economic recession hit the United States, the payment bank went bankrupt, and this company ran into financial difficulties. Moreover, in 1894, Nikola Tesla invented alternating current electricity, which has minimal power loss even in long-distance transmission, making Love's business plan obsolete. Eventually, Love had to stop the business in 1910, leaving only a disgraceful huge pit called Love Canal, which is 1.6 km long, 14 m wide, and 3–12 m deep.

For decades afterwards, Love Canal was left abandoned as an eyesore, and young children in this area enjoyed swimming in small pits made here and there in the canal. In the 1940s, a chemical factory called Hooker Chemical found out that the canal floor was made of mud and took it over, judging that it was very suitable for dumping its industrial waste. In 1942, Hooker Chemical applied to the New York

S. S. Park, *Eco-disasters*, https://doi.org/10.1007/978-981-96-8667-4_20

State government to use this place as an industrial waste landfill and was approved. At that time, Hooker Chemical was producing pesticides, herbicides, and organic solvents in nearby areas.

Hooker Chemical installed a concrete impermeable layer on the canal floor and began to landfill industrial waste in 55-gallon (208ℓ) iron drums, dumping a whopping 22,000 tons of 248 types of toxic chemicals here between 1942 and 1950. This included pesticides such as lindane and DDT, various organic solvents, PCBs, dioxins, heavy metals, etc. In addition, garbage discharged from Niagara City and harmful waste generated from the Manhattan Project, where the US Army was developing nuclear weapons during World War II, were also buried here.[1]

In 1953, Hooker Chemical covered the area with clay and concrete, then donated it to the Niagara Falls Board of Education for $1, along with the surrounding land. The Board of Education built an elementary school there in 1955, and hundreds of homes were constructed in the vicinity. However, over time, rainwater seeped into the buried waste, and the metal drums began to rust.

In the early 1970s, strange chemical substances began to emerge from the elementary school playground built on the landfill. Students found it fascinating to throw stones at these substances, which would emit smoke and corrode. At that time, there was a lack of environmental awareness, so the school authorities and citizens were indifferent despite knowing about these occurrences. Additionally, unusual substances occasionally leaked into the basements of nearby homes, and sewer pipes corroded from a black liquid, yet little attention was paid to these issues.

Residents of this area often suffered from skin diseases and headaches, and the miscarriage rate was higher than in other areas. However, after a major flood in 1976, the street trees and garden flowers in this area began to wither and die. Toxic chemicals were detected in large quantities in the pond where they enjoyed swimming, and moisture containing toxic substances seeped from the soil to the surface. Also, many residents complained of physical pain. In 1977, the authorities who investigated this area confirmed that the groundwater was severely contaminated with toxic chemicals. Nevertheless, the school and city authorities took no significant action.

Meanwhile, a newspaper reporter named Michael Brown began to investigate and report on the expected problems in the Love Canal area, considering a series of events taking place in 1976. In 1978, through a local newspaper called the "Niagara Gazette," Brown reported that the toxic chemicals buried in Love Canal were causing serious damage to the residents' health. Upon reading this article, a parent named Lois Gibbs believed that her son's chronic asthma, kidney disease, and epilepsy were due to the toxic chemicals buried under the school. She requested a transfer from the school authorities, but was denied. Her son experienced a continual decrease in white blood cells and even suffered from strokes after enrolling in this school. Gibbs decided to petition for the closure of the school and visited the homes of students daily to investigate their health conditions. Surprisingly, most of the children were suffering from various diseases. She formed a residents' committee and sent a petition to President Carter stating, "We do not want our children to grow up in an environment

[1] How to Survive in America the Poisoned, Lewis Regenstein, Acropolis Books, 1986.

where they suffer from diseases and their future is uncertain."[2] Some angry residents held environmental protection officials hostage for six hours to make their plight known and assert their demands. The residents informed various media outlets about the incident and even protested at political rallies.

As the impact of the incident grew, the New York State Health Department conducted an epidemiological investigation of the area. The investigation revealed that the miscarriage rate among residents in this area was four times higher compared to other areas. It was also uncovered that out of 16 children born between 1973 and 1978, nine suffered from severe congenital defects such as mental retardation or heart and kidney diseases. Recognizing the severity of the problem, the Environmental Protection Agency declared this area an environmental disaster area for the first time in U.S. history in August 1978 and ordered the 238 households living there to leave immediately. The problematic elementary school was closed, and residents' compensation lawsuits were filed. The residents claimed billions of dollars in compensation, but there was no responsible party to pay it.

The Environmental Protection Agency continued to investigate the extent and severity of the contaminated soil and added 800 households to the disaster area in May 1980. In this disaster area, equivalent to 160,000 m^2, houses and schools were demolished, and to prevent the evaporation or leaching of harmful substances and block rainfall, a plastic cover was placed over the area, and about 0.5 m of soil was piled on top and grass was planted. A 2.5 m high fence was also built to prevent access by people and animals.

In response to this incident, the U.S. federal government enacted the Superfund law in December 1980, establishing a federal fund worth $16 billion over the next five years to promptly address issues arising from polluted areas like Love Canal. Before this law was enacted, under the U.S. environmental law system, when a hazardous area was identified, the perpetrator and the victim determined the compensation through litigation, which required more than a year until the final judgment. The basic content of the Superfund law is that the Environmental Protection Agency can quickly compensate for hazardous areas with government funds first and then find the perpetrator afterwards and demand up to three times the cost of compensation. The term 'Superfund' originated from the meaning that it transcends the general compensation law procedure, and its official name is 'Comprehensive Environmental Response, Compensation, and Liability Act (CERCLA)'. By the statute of limitations of this law, September 30, 1985, the Environmental Protection Agency found 27,660 polluted areas and carried out hazardous substance removal work in 507 highly dangerous areas. This law, which was intended to be temporarily applied for five years, was revised and extended under the name of the Superfund Amendments and Reauthorization Act by the Reagan administration in October 1986, and an additional $8.5 billion fund was added. By 1990, the Environmental Protection Agency had found 26,000 polluted areas, 1,224 of which were classified as severely polluted areas.

---

[2] Love Canal: My Story, Lois Gibbs, Evergreen, 1982.

The U.S. federal government spent more than $100 million to clean up this area, but for a long time, it remained a ghost town where no one could enter. Some areas with less severe contamination, thanks to ongoing cleanup efforts, were able to be re-inhabited only after ten years of closure. In September 1992, a new housing complex was redeveloped here, and 31 households began to move in for the first time. The Love Canal incident was one of the worst environmental disasters that stirred up the entire United States at the time. In particular, it is a good example of showing that indifference and ignorance about the environment lead to tremendous disasters in the future. The fact that a woman who had lived here for a long time said, "We knew that where we live was an industrial waste landfill, but we didn't know it would make our family like this," shows how ignorant people were about the environment at the time.

The responsibility for compensation for damages eventually fell on Hooker Chemical and the Niagara City Board of Education. The land donated for the education of future generations ironically made them sick and caused deaths, all due to environmental ignorance. Many people still think that anything buried in the ground disappears. However, we have to remember that it comes back to us as a boomerang with tremendous life-killing power.

Chapter 21
Darling River Algal Bloom

The Darling River basin, located in southeastern Australia, has flat grasslands that are well-suited for livestock farming. In December 1991, an incident occurred where about 1,600 cattle and sheep that were grazing in the pastures of the Darling River basin died en masse.[1] An investigation into the cause of death revealed a problem with the river water the livestock had consumed. At the time, the Darling River was suffering from a severe drought that had lasted over a year, resulting in a shortage of water and a severe state of eutrophication. High water temperatures and intense sunlight led to a rapid growth of algae.

Generally, diatoms or green algae grow in the water during low-temperature periods when the water temperature is low, and blue-green algae (Blue Green Algae, Cyanobacteria) grow in the water during high-temperature periods when the water temperature is high. Particularly when the water temperature rises rapidly due to abundant nutrients and strong sunlight, blue-green algae increase explosively. This blue-green algae produces substances that are highly toxic to many animals, including humans. The livestock of the Darling River died from drinking water contaminated with the toxic substances of the blue-green algae.

Eutrophication, the root cause of algal blooms, can be compared to human obesity. While a certain amount of nutrients is essential for the human body, too much can lead to obesity and harm health. Similarly, if there are appropriate nutrients in the water, aquatic plants and animals can live in harmony and balance, but if there are too many nutrients, plant plankton overbreed, increasing organic matter and clouding the water. Organic matter can induce oxygen depletion, killing many aquatic animals including fish, and producing toxic substances like ammonia, causing bad smells and even worsening the taste of the water, making it unable to use as a source of drinking water. Also, the non-biodegradable organic matter produced by algae can produce disinfection by-products that cause cancer and neurological abnormalities during the water purification process. Especially, blue-green algae, which secrete

[1] https://en.wikipedia.org/wiki/1991_Darling_River_cyanobacterial_bloom.

S. S. Park, *Eco-disasters*, https://doi.org/10.1007/978-981-96-8667-4_21

toxic substances, are the most dangerous among freshwater plant plankton. Blue-green algae can be distinguished because they look similar to green algae in color, but they mostly float on the water surface, forming scum or appearing as if green powder has been sprinkled, and when the wind blows, they gather at the water's edge. Drinking water from areas where blue-green algae are concentrated can lead to poisoning accidents.

The toxins in blue-green algae include hepatotoxic substances that cause fatal harm to the liver, neurotoxic substances that paralyze nerves, and glycolipids that cause stomach upset or allergic dermatitis. A liver toxin called Microcystin, which is so toxic that it can kill 50 experimental mice with just 1 mg, can be fatal to humans even in small amounts. A neurotoxin called Anatoxin can paralyze nerves in a short time and can be fatal. Acute poisoning, where a cow drinking water from a lakeside suddenly falls down and convulses or dies within a few minutes, is due to this neurotoxin. Unlike liver toxicity, where lethal effects appear a few hours later, the effects of neurotoxicity appear immediately, and if it does not lead to death, it recovers without major sequelae over time.

Damage cases due to cyanobacterial toxins have been frequently reported. The most common cases are the deaths of grazing livestock or wildlife. In large pastures in Australia and the United States, most cases involve cows freely grazing and then going to lakes or reservoirs to drink water and die.

Studies have shown that salmon that consume cyanobacteria for a long period of time shrinks and eventually dies because this toxic substance also damages the liver of fish. However, carp and goldfish are fish that live well even after eating cyanobacteria, and are known to be resistant to cyanobacterial toxins.

Damage caused by cyanobacterial toxic substances has been frequently reported. The deaths of grazing livestock or wildlife are the most common cases. In most cases, cows visit lakes or reservoirs to drink water and die while freely grazing on large ranches in Australia and the United States. Studies have shown that salmon that consume cyanobacteria for a long period of time shrinks and eventually dies because this toxic substance also damages the liver of fish. However, carp and crucian carp are fish that survive well even after eating cyanobacteria, and are known to have resistance to cyanobacteria.

Toxic substances of blue-green algae are filtered out of tap water purification plants, so accidents that can lead to death by humans are rare. However, if water that has generated a large amount of blue-green algae is used outdoors or if a problem occurs in the water purification plant, it can lead to a serious accident. In 1976, 62% of 8,000 residents in Sewickley, Pennsylvania, USA, who drank tap water produced by purifying the water from reservoirs contaminated with blue-green algae, developed diarrhea and gastroenteritis. In 1979, more than 150 residents in Palm Island, Queensland, Australia, experienced abnormal symptoms in their stomachs, liver, and kidneys after drinking tap water from a reservoir where blue-green algae flourished. In addition, in 1988, more than 2,000 residents who used water from reservoirs where blue-green algae flourished suffered severe abdominal pain, killing 88 people in 42 days, most of them children. Boiled water was used for drinking water here, so it was not a cause of bacteria or viruses. This incident demonstrates that blue-green algae toxic substances do not decompose even at high temperatures.

The most damaging incident due to cyanobacteria occurred in 1996 in the Caruaru region of Brazil, where 131 people showed symptoms of poisoning such as visual impairment, nausea, and vomiting after using reservoir water containing cyanobacterial toxins for kidney dialysis, and 55 of them lost their lives. This incident occurred because the water supplier transported water taken from a source where cyanobacteria had occurred to a water tank without properly treating it, and merely disinfected it with chlorine. The water was then supplied to a specialized kidney dialysis hospital, and the toxins were directly introduced into the bodies of patients who used this water for treatment. At the time, the water used for dialysis and human tissues detected microcystin, a liver toxic substance.

Other cases of damage caused by cyanobacterial toxicity are frequently reported. Especially in developing countries, a lot of damage is occurring because they use the water from rivers or lakes where cyanobacteria live as drinking water without properly purifying it. If you consume a low concentration of liver toxic substances (microcystin) for a long time, it leads to liver cancer. In one region of China, the incidence of liver cancer was found to be much higher than in other regions, and as a result of investigating the cause, it was attributed to cyanobacteria living in the lake water used for drinking.

The key role in causing algal blooms in rivers and lakes is phosphorus, a fertilizer component of plants. The increase in phosphorus in water accelerated in the late 20th century as the use of chemical fertilizers surged. To increase the amount of food production per unit area, it is necessary to supply fertilizers such as phosphorus and nitrogen, and in the past, compost was made from human or livestock phosphorus or manure and spread on the soil. As chemical fertilizers were invented and industrialization progressed, phosphorite was mined, allowing farmers to increase food productivity by spraying phosphorus more efficient than compost into the soil. However, phosphorus and manure were not recycled to the soil and went to rivers and lakes. Not only that, chemical fertilizers spread on the soil also go into rivers and lakes when washed by rainwater.

Phosphorus fertilizer was invented by British farmer John B. Lawes.[2] He lost his father at the age of eight and was left to take care of the farm left as an inheritance. At a young age, he observed changes in crop production according to manure spreading while farming alone. Then, in 1842, he developed a groundbreaking method of treating phosphorus with sulfuric acid to make fertilizer. He patented this and started the chemical fertilizer industry for the first time in human history. For this pioneering achievement, he became a Fellow of the Royal Society in 1854 and awarded the Royal Medal in 1867. The invention of phosphorus fertilizer by British farmer John Lawes remains a great achievement of civilization that brought about an increase in human food production.

Another factor that increases phosphorus in rivers and lakes is synthetic detergents. Synthetic detergents began to be developed to replace animal and vegetable fats,

---

[2] https://en.wikipedia.org/wiki/John_Bennet_Lawes.

the raw materials for soap, which were used to manufacture cannonballs during World War I, and after World War II, they were used in general households as their synthesis methods and cleaning effects improved. However, synthetic detergents are toxic to aquatic life, do not easily decompose in water, and contain a large amount of phosphorus. The phosphorus contained in synthetic detergents and fertilizers is all extracted from mines.

As the use of chemical fertilizers and synthetic detergents increased, serious eutrophication appeared in rivers and lakes, mainly in advanced industrial countries, and various efforts were made to overcome this. One of the representative examples is Lake Geneva in Switzerland. Lake Geneva, which suffered severe eutrophication due to domestic sewage and synthetic detergents, was revived through international cooperation.

In the early 1950s, after the end of World War II, Switzerland and France, located in the Lake Geneva basin, actively carried out post-war recovery, rebuilt destroyed cities, and agriculture and industry began to be revitalized. However, during this vibrant reconstruction activity, the clear water of Lake Geneva, the largest in Switzerland, gradually rotted, and by the late 1950s, it had turned into a dead lake where life could no longer survive, not to mention the stench.

In 1962, Switzerland and France began to investigate pollution sources in the basin and install sewage treatment plants at major pollution sources first, through a bilateral agreement to prevent pollution of Lake Geneva.

Over the next 20 years, about 120 sewage treatment plants were built in the Lake Geneva basin, and the basin was modified to divert polluted rivers below the lake, greatly improving the water quality. Thanks to these efforts, Lake Geneva revived as a clear and beautiful lake from the late 1970s.

In some developed countries, algal blooms have been reduced through measures to prevent eutrophication, such as the case of Lake Geneva. However, in developing countries like China and India, where the sewage treatment rate is low, the degree is very serious. Even in countries with relatively high sewage treatment rates such as the United States, the United Kingdom, and Japan, algal blooms have recently become more frequent due to drought and heat waves.

The best way to prevent algal blooms is to reduce nutrients. However, if drought and heat waves continue, even if the inflow of nutrients is reduced, evaporation and concentration can create conditions favorable for the proliferation of cyanobacteria in rivers and lakes. Therefore, with regard to the algae phenomenon, two measures are taken together: prevention and post-occurrence treatment. The most widely used method since the past as a post-occurrence treatment measure is to collect algae on the surface and use it as fertilizer. Cyanobacteria rise to the surface and form clumps, so it is relatively easy to collect them using a boat. The boat sucks in the algae, filters and dries it to make fertilizer. When sprayed into the soil, algae fertilizers have advantages over chemical fertilizers because they decompose slowly and supply nutrients.

Recently, more groundbreaking research on green algae is being conducted in developed countries. It is about producing bioenergy and bioplastics from green algae. Green algae is the subject of interest for many bioenergy experts because it shows a production volume 20–30 times higher than the highest yield per unit area among the bioenergy crops used so far, such as palm oil and coconuts. Not only that, but it also has the advantage of easy production even when the water quality is poor. Research on producing bioplastics from green algae is also active. In the United States and Japan, they are currently producing bioplastic household goods. Compared to regular plastics, it is very environmentally friendly and has many advantages, such as easy decomposition after use and reduction of oil consumption. In addition, various new material developments, including cosmetic ingredients from green algae, are also underway.

The Darling River algae bloom shows us that natural water can turn into deadly poison due to drought and heat waves in eutrophic conditions. Similar incidents can occur at any time because algal blooms often occur in rivers and lakes where people live and livestock activities take place, blue-green algae often occur. In particular, special countermeasures are needed in places where water for living and industry is obtained from rivers and lakes with frequent droughts and heatwaves.

Chapter 22
Guatemala Shellfish Poisoning

In July and August of 1987, 187 residents who ate soup made from sea clams in Champerico on the Pacific coast of Guatemala suffered from food poisoning (Paralytic Shellfish Poisoning), resulting in 26 deaths.[1] Children under the age of six showed a high mortality rate of 50%, while adults over 18 had a mortality rate of 7%. The cause of their deaths was not a common waterborne disease caused by seafood consumption, but a toxic substance contained in the clams, which does not decompose even when boiled. This toxic substance was not from wastewater flowing from the land, but was produced by algae living in the sea. At the time of the incident, the sea was showing a red tide phenomenon, where algae proliferate explosively and turn the water a dark red color.

Red tide is a phenomenon that occurs when there are abundant nutrients such as nitrogen and phosphorus in seawater and there is a lot of sunlight, especially in warm harbors where the water flow is stagnant. There are about 4,000 species of algae living in the sea, and it is estimated that about 2% of them secrete toxic substances. When a red tide occurs, eating clams or fish that have consumed toxic algae can lead to death. Additionally, contact with seawater can cause symptoms such as skin spots, fatigue, and nausea.

Similar cases can be found around the world. For example, in 1987 when 157 residents of Prince Edward Island in Canada suffered from abdominal pain after eating mussels, resulting in four deaths. In the United States, red tides occurred in North Carolina in 1995 and Maryland in 1996, causing millions of fish to die en masse. The red tide organism known as Pfiesteria that occurred at this time is known to secrete a toxic substance a thousand times stronger than cyanide. Therefore, warnings against shellfish consumption are often posted on coasts where red tide is at risk.

The toxic substances produced by red tide organisms are of two types: those that cause nerve paralysis and those that cause digestive disorders. Nerve paralysis toxicity causes muscle paralysis, headaches, dizziness, respiratory difficulties,

[1] https://pubmed.ncbi.nlm.nih.gov/2316796/

S. S. Park, *Eco-disasters*, https://doi.org/10.1007/978-981-96-8667-4_22

suffocation, and can lead to death in severe cases. Digestive disorder toxicity causes abdominal pain and diarrhea. Some red tide organisms secrete toxic substances that irritate the eyes, nose, throat, and lungs, causing symptoms similar to asthma or pulmonary emphysema. These substances are mostly consumed through red tide organisms that remain in the intestines of shellfish, and once these toxic substances enter the body, they do not easily leave.

When the rapidly increased red tide organisms die, the dissolved oxygen in the seawater decreases rapidly and hydrogen sulfide, ammonia, and methane gas are generated, causing most of the species living in the red tide area to die en masse. It is particularly fatal to fish and shellfish that breathe oxygen in densely populated fish farms. When a red tide occurs, mammals such as whales, seals, dolphins, and manatees, and birds like waterfowl, die from toxic substances, but fish and shellfish breathing with gills are more often suffocated than toxic substances because oxygen is not delivered due to the attachment of the gills of red algae.

There are various types of algae that cause red tides, but dinoflagellates, diatoms, and flagellates are the main red tide organisms. These are usually single-celled organisms with a diameter of 20–40 μm, most of which are red, but sometimes they are yellow, brown, green, and there are also luminescent species that emit light at night.

The routes of entry for substances causing red tide are very diverse. Domestic sewage and livestock waste discharged from the land account for the largest portion of the nutrients in coastal seawater. Most of the waste water is now treated, but most of the organic matter is removed and a considerable amount of nutrients are discharged. In addition, excrement and feed from coastal fish farms are also significant causes of red tide. Futhermore, fertilizers sprayed on farmland are washed away by rainwater and flow into the sea, contributing to the occurrence of red tide.

Once a red tide occurs, there are methods such as killing red tide organisms by spraying chemical substances, directly filtering and removing red tide organisms from the sea surface, and using clay or polymer coagulants to adsorb and coagulate red tide organisms and cause them to settle. However, these methods are uneconomical and have low effectiveness because they are carried out in the open sea. Therefore, most countries leave it as it is until nature recovers on its own after a red tide occurs.

In order to reduce the damage from red tide, proactive preventive measures must be prepared above all. It is necessary to thoroughly manage pollution sources including sewage in coastal areas and devise ways to discharge sewage outside the closed sea area. Additionally, by regulating the use of fertilizers, the nutrients flowing into the sea with rainwater from farmland should be reduced, and during the red tide occurrence period, at least, a plan to remove nutrients by highly treating sewage from coastal cities should be considered. Coastal areas sensitive to red tides should be managed with stricter environmental standards.

The frequency of red tide occurrences is continuously increasing worldwide since the 1980s. Major occurrence areas are Asia (Korea, Japan, China, India, Philippines, Thailand, etc.), Central and South America (Guatemala, Venezuela, Argentina, Brazil, Peru, Chile, etc.), Europe (Italy, Spain, France, UK, etc.), and the United States, Australia, New Zealand, etc. Recently, red tides have also been reported in relatively low sea temperature areas such as Canada, Scotland, Denmark, Norway, and Russia.

The reason for the increasing frequency of red tides worldwide is because the concentration of nutrients in seawater continues to increase and the water temperature is gradually rising. Red tides occur for the same reasons as algal blooms in fresh water, but nitrogen plays a more important role than phosphorus (Refer to Chap. 21). This is because seawater is relatively rich in phosphorus compared to nitrogen. Therefore, when strong sunlight continues and the water temperature is appropriate, the growth of red tide organisms is determined by the nitrogen concentration. And a large portion of the nitrogen in coastal seawater is nitrogen fertilizer spread in the watershed.

Nitrogen fertilizer was made possible by the groundbreaking invention of Fritz Haber, a professor at the Karlsruhe Institute of Technology in Germany, in 1909.[2] Nitrogen exists indefinitely in the air in gas form, but plants cannot use it as fertilizer. Professor Haber invented a process (Haber Process) that combines nitrogen in the air with hydrogen under high temperature and pressure to make ammonia. When this produced ammonia is spread on the soil, it becomes nitrate by microbes and can be used by plants. This invention brought about tremendous changes in human history. Nitrogen fertilizer factories were built all over the world and agricultural production increased dramatically. Half of the food that humans around the world eat today has been made possible by the nitrogen fertilizer process invented by Haber. Haber won the Nobel Prize in Chemistry in 1918 for this invention.

However, Haber led the development of chemical weapons for Germany during World War I and faced a tragic family history. He made various poisonous gases including chlorine gas and developed gas masks, making it possible for chemical warfare to kill the enemy without harming our troops. The ammonia synthesis process he invented was also used to manufacture bombs. His wife Clara, who was also a chemist and well aware of Haber's research, Clara Immerwahr vehemently opposed the inhumane and brutal research into chemical warfare. Despite his wife's objections, Haber passionately developed chemical weapons capable of killing humans. In the end, Clara, in an attempt to change her husband's mind, left behind her 13-year-old son and committed suicide with a pistol in 1915, at the age of 44.

Even after his wife's suicide, Haber continued to develop chemical weapons, stating that "scientists should contribute to humanity in times of peace, but must be loyal to the nation in times of war" He died in 1934 at the age of 67. What was left of him was not a great contribution to humanity, but rather a disgrace as the Father of Modern Chemical Warfare. Haber's great achievements and tragic family history once again make us reflect on the dual nature of modern science. The nitrogen fertilizer production process he invented also saved countless people from starvation and provided abundant food, but it left an inconvenient by-product, the recurring algal blooms.

---

[2] https://en.wikipedia.org/wiki/fritz_haber

Chapter 23
Times Beach Dioxin

In a small rural town called Times Beach in central Missouri, USA, oil was sprayed on the roads for several years from 1971 to reduce dust from unpaved roads. But then after, strange things began to happen in this town. Birds flying in the sky fell to the ground, and cats and dogs died. Residents also felt pain without any particular reason, developed lung cancer, and pregnant women had miscarriages. Moreover, there were many residents who suffered from diseases in the liver, kidneys, larynx, or developed cancer.[1]

The act of spraying oil to reduce dust was not only on the roads, but also took place in ranches where horses were raised and trained. In May 1971, after oil was sprayed in the horse training field of the Shenandoah Ranch in Moscow Mill, a town near Times Beach, 62 horses died. And the ranch owner's two young daughters complained of severe pain for unknown reason. Similar damage appeared at the Timberline Ranch in nearby Jefferson City since oil was sprayed. To find out the cause, they requested an investigation from local doctors and informed the Centers for Disease Control and Prevention (CDC) under the Federal Health Department of this situation.

The CDC visited the ranch in August 1971 to investigate the cause. However, the first report that released a year later only contained the phrase 'cause unknown'. It was not until August 1974, two years later, that the CDC announced that the cause of the strange disease phenomena that had occurred at various ranches in Missouri was dioxin. The oil sprayed on the unpaved roads and ranches contained a large amount of dioxin, which caused diseases in humans and animals. Dioxin is a general term for chemical substances in the Dibenzo p-Dioxins group, especially 2,3,7,8,-TCDD (2,3,7,8-Tetrachlorodibenzo-p-dioxin) is highly toxic and causes various diseases including cancer in the human body even in extremely small amounts. This substance is produced as a by-product in the manufacturing process of herbicides (2,4,5-T) and is also created in the process of incinerating garbage. Dioxin was the cause of the Agent Orange incident in Vietnam and the Seveso incident in Italy, and has often

[1] https://www.epa.gov/mo/town-flood-and-superfund-looking-back-times-beach-disaster-nearly-40-years-later

S. S. Park, *Eco-disasters*, https://doi.org/10.1007/978-981-96-8667-4_23

been a problem in environmental disasters occurring in industrialized countries. The residents of Times Beach were horrified to learn that this highly toxic substance had been sprayed on the roads of their town.

Soon after, it was revealed that the company that had sprayed oil on the roads and ranches was a waste oil collection and regeneration company run by a man named Russel M. Bliss. It was revealed. Russell Bliss was commissioned to dispose of industrial waste generated by a pharmaceutical company called Nepacco, located in Verona, Missouri, and since 1971, he has been collecting 18,500 gallons (70,300 L) of high-concentration dioxin-containing material annually, mixing it with waste oil, and spraying it all over Missouri.

From 1970 to 1972, the Nepacco pharmaceutical company manufactured a substance called Hexachlorophene. Initially, Nepacco transported the industrial waste generated in the process of manufacturing this substance to a hazardous waste incinerator in Louisiana for disposal, but due to high costs, they commissioned a petrochemical company located in nearby St. Louis for disposal. However, this company entrusted the disposal to Russell Bliss, who was in the waste oil collection business. Russell Bliss, who also raised horses, sprayed waste oil to reduce dust from his ranch. This method was very effective, and the effect lasted for several months after one spray. As the effect became known, requests came in from all over the place, and eventually, waste oil was sprayed on nearby roads and horse training grounds.

In the early stages of the investigation into the degree of contamination and damage in the area where waste oil containing dioxin was sprayed, the seriousness of the problem was not highlighted. At that time, the chemical properties and toxicity of dioxin were not scientifically established, and there were no legal mechanisms to regulate it in the United States. Experts initially thought that dioxin would decompose and disappear from the soil within a few years. However, after investigating the contaminated area for several years, it was found that dioxin does not decompose and remains in the environment for a long time. Also, there was no way to decompose this substance into harmless substances except by burning it at temperatures above 2,000 °C, so there was virtually no way to restore the contaminated area in a short period of time.

When the situation reached this point, the problem was not limited to the waste lubricating oil sprayed on Times Beach and the ranch. The U.S. Environmental Protection Agency had to track where all the industrial waste from the Nepacco pharmaceutical company had gone. In 1980, the Environmental Protection Agency found a place where Nepacco had dumped a large amount of industrial waste ten years ago (before commissioning Russell Bliss) and discussed post-measures along with a full-scale damage investigation. Eventually, in 1983, the federal government bought all areas in Missouri contaminated with dioxin. They forcibly relocated over 22,000 residents. All places where industrial waste was dumped, including Times Beach, became uninhabitable lands of death.

This incident was an absurd environmental disaster caused by illegal dumping of industrial waste and human ignorance. At that time, industrial waste treatment technology was not developed and the management system was inadequate, so illegal dumping was carried out everywhere. Industries thought that they just had to hand over their waste in an appropriate way, and they did not care where and how the waste disposal workers disposed of it. Not only that, it cost a lot to dispose of waste safely, so it was a business target for criminal organizations like the Mafia to illegally dump it at a cheap price, and it also became a factor in corruption politics.

In fact, in Alabama, USA, an incident occurred where an Environmental Protection Agency official investigating an illegal hazardous waste landfill was shot by the mafia. On March 27, 1983, three Atlanta Environmental Protection Agency employees discovered a huge pile of iron drums filled with toxic chemicals while searching for industrial waste dumped in the forest. The iron drums were rusted and toxic substances were seeping out. Suddenly, gunshots began to ring out around them. Using the hazardous material drums as shields, they hid their bodies and fled from there. Similar incidents occurred in Philadelphia that same year. It was common for people investigating illegal hazardous waste landfills to be shot, severely beaten, or attacked by stray dogs. In Seattle, bombs were dropped to hide piles of hazardous materials, and in Chicago, death threats were made to those who provided information about illegal dumping areas to the Environmental Protection Agency. In 1984, Los Angeles launched the Toxic Waste Strike Force to eradicate illegal dumping, sending all senior executives of 12 local companies to prison. And the Los Angeles court sent a message across the United States that illegal dumping of hazardous waste is the worst crime against local residents. Some companies even used one of their factories as a waste dumping ground.

Vertac, a chemical factory in Jacksonville, Arkansas, closed the factory, leaving tens of thousands of drums of hazardous waste containing high concentrations of dioxin. Toxic substances seeped from rusty drums, contaminated soil and groundwater rolls, and flowed into rivers through sewers. Also, there was sludge concentrated with dioxin piled up to a height of 6 m. This was enough to cover 20 football fields to a depth of 0.9 m. Subsequently, the local residents showed high rates of cancer incidence, birth defects, and infant mortality, and 65 residents who were proven to have suffered direct damage received relocation costs from this company and left the area. Illegal dumping of industrial waste is a problem occurring not only in the United States but also in most industrialized countries. The U.S. Environmental Protection Agency reported in the early 1980s that over 90% of industrial waste in the United States was illegally dumped, and there are tens of thousands of such dumping sites. Industrial waste is passed from the emitting company to criminal organizations, illegally dumped, and ultimately the damage returns to the residents. Then, with the taxes paid by the residents, a huge cost is spent to carry out the cleanup project in the illegal dumping area, and this vicious cycle is still being repeated in many areas today.

In 1999, a lush 419-acre (1.7 km^2) US Route 66 State Park was opened in the Times Beach area. It was named after its proximity to the highway (US Route 66) connecting Chicago and Los Angeles. Times Beach, a small rural town, disappeared due to absurd environmental ignorance and became a roadside park after a long soil purification process. The U.S. Environmental Protection Agency revisited the site in June 2012, collected and analyzed soil, and reported in November of the same year that "the soil of this park poses no particular risk to the health of workers and visitors here." It suggests that even after restoration, places that were contaminated with hazardous substances should be re-examined for health risks in case of human contact.

Chapter 24
Gulf War Eco-Terror

A significant amount of the world's oil is produced in countries such as Kuwait, Saudi Arabia, Iraq, the United Arab Emirates, Iran, Oman, and Qatar, all located along the coast of the Persian Gulf. On August 2, 1990, Iraq forcibly occupied Kuwait. After several months of negotiations at the UN, a counterattack led by the United States began on January 17, 1991, and by March 3, Kuwait was finally liberated from Iraq. During the liberation of Kuwait, fierce battles took place between Iraq and the UN forces, resulting in significant property and human casualties, as well as a malicious environmental disaster of the twentieth century.

On January 25, 1991, Iraq bombed Kuwait's oil terminals located in the Persian Gulf, causing over a million tons of crude oil to spill into the sea.[1] The Iraqi army also blew up over 800 oil wells and completely destroyed related facilities such as pipelines and storage tanks during their withdrawal from Kuwait. Of the wells that were blown up, over 600 were left in flames, burning up to 4.6 million barrels of oil daily.

This brutal environmental terrorism was a deliberate act by Iraq to contaminate the Persian Gulf with oil and shake the global oil market, using it as a means of threatening Western countries. The amount of oil spilled at the time was greater than any oil tanker accident or offshore oil leak in the world to date. The oil flowing from the destroyed terminals covered over 450 km of coastline stretching from Kuwait to Saudi Arabia.

The Persian Gulf is a closed sea, so most of the spilled oil remained within the gulf, but there was no way to remove it during the war. If oil spills into the sea, damage can be greatly reduced by taking preventive measures such as installing oil fences and spraying absorbents or dispersants in the early stages of the accident. However, no measures could be taken for the oil spill in the Persian Gulf because it occurred during the war. Therefore, the oil spill in the Persian Gulf was the worst

[1] https://en.wikipedia.org/wiki/gulf_war_oil_spill

S. S. Park, *Eco-disasters*, https://doi.org/10.1007/978-981-96-8667-4_24

marine environmental disaster in history, with a huge amount of oil spilled into a closed sea without any preventive measures taken.

The coast of the Persian Gulf is widely covered with coral reefs and sandy beaches, and especially, a unique coastal wetland called Sabkha, a mixture of sand and pebbles, is formed, where a very diverse range of organisms lived. There were a variety of edible fish, invertebrates, and rare flora and fauna living only here due to the overgrown weeds, and Sabka's diverse biology and high productivity have served as a good source of food for the Persian Gulf marine ecosystem. However, after the Gulf War, oil lumps were scattered on coral reefs and sandy beaches, and birds that could not find food also died in droves.

Saudi Arabia, Bahrain, Qatar, and the United Arab Emirates, located on the coast of the Persian Gulf, were using more than 90% of their drinking water by desalinating the seawater of the Persian Gulf due to a lack of water resources. However, due to this accident, desalination plants built in Jubail, Dammam, and Barad could no longer produce drinking water and were closed, and the residents here could not even get drinking water.

In the Gulf War, birds died and street trees withered due to air pollution caused by burning oil wells in the Persian Gulf, and people were suffocating. At that time, experts were worried that the huge amount of black smoke emitted from the burning oil wells would rise high into the sky and ride the jet stream. If the smoke rises to 45,000 feet (about 13,700 m) above the ground and joins the jet stream, soot and petroleum chemicals will spread all over the world. In this case, it can have a serious impact not only on the pollution of the upper troposphere but also on the global climate. However, fortunately, the smoke emitted at this time only reached up to 22,000 feet (about 6,700 m) above the ground. According to the surveyed data, the soot emitted at this time was detected in the snow that fell on the Himalayas, but most of it was found to have affected the surrounding countries of the Persian Gulf. The task of catching the flames of more than 600 oil wells engulfed in flames was the biggest problem after the recovery of Kuwait. Special firefighters dispatched from all over the world made their best efforts, but it took two years to catch the flames of the oil wells. Due to the smoke generated at this time, residents of Kuwait and Saudi Arabia had to suffer from respiratory diseases, and acid rain even fell in nearby Middle Eastern countries.

A large amount of oil from destroyed oil fields flooded the ground, polluting the soil and covering rivers, lakes, and all living things on the ground. Oil permeated deep into the soil, causing such a major change in the terrestrial ecosystem. The total amount of oil leaked is estimated to be approximately 6~8 million barrels (95,340~1.27 billion liters). The damage to soil pollution caused by oil flooding was less direct than that of air pollution or marine pollution that occurred at the time, but it lasted the longest.

Most of the environmental disasters that have occurred on Earth so far have been due to human carelessness and ignorance about the environment. However, the oil well explosion during the Gulf War was an act of environmental terrorism intended to deliberately cause a disaster and use it as a means of war. At the time, a foreign newspaper described the Gulf War as a 'war waged by taking the Earth hostage', which caused a horrific environmental catastrophe.

Looking at human history, war has always been accompanied by brutal environmental disasters. For thousands of years, wars have always been conducted in a way that devastates the enemy's land. The occupying forces burned the houses and fields of the enemy country, looted livestock and grain, and poisoned the wells. Killings, looting, arson, and other acts leading to environmental disasters are still seen in all wars. Recently, wars have been more thoroughly destroying everything necessary for survival, such as roads, railways, power plants, and hospitals. The Gulf War caused a more horrific environmental disaster than any previous war by destroying the numerous oil wells and oil transportation facilities concentrated in this region. And the damage has been left very extensive and unrecoverable for a long time.

Wars occurring in today's industrial society can cause tremendous damage because they can destroy hazardous facilities such as nuclear power plants and toxic chemical manufacturing plants. In this case, even if weapons of mass destruction such as nuclear or biological and chemical weapons are not used, it can lead to a tragedy beyond imagination. Of course, it is best to prevent war in advance, but in order to minimize possible environmental disasters caused by it, countermeasures such as urban planning, industrial facility deployment, and emergency systems should be prepared taking this into account. Also, it is a way to a sustainable future for the Earth and humanity to establish a second Geneva Convention that prohibits acts of destroying facilities that could lead to large-scale environmental disasters during war.

Part V
Nuclear and Industrial Accidents

Chapter 25
Chernobyl Nuclear Disaster

On the morning of April 28, 1986, a worker at a nuclear power plant in Sweden was checking the instrument panel of the reactor as usual. However, something very strange was discovered that day. Despite no abnormalities in the reactor, high levels of radiation were observed throughout the power plant. The worker also found an unknown radioactive substance on his shoes.

He immediately inquired about any abnormalities at nuclear power plants and radiation observatories across Sweden, and found that the same question was being raised in other areas. When the observed substance was analyzed in the laboratory, it turned out to be krypton, iodine, and cesium, which are produced during nuclear fission in the reactor. Also, wind direction analysis suggested that these substances had moved from the Soviet Union. That evening, Moscow Broadcasting reported an accident at a nuclear power plant located in Chernobyl, Ukraine.[1]

The Chernobyl nuclear power plant accident occurred due to a mistake by a worker conducting a turbine shutdown experiment just three days before it was known to the Western world. In a nuclear power plant, the heat generated in the reactor during nuclear fission is used to boil water, and the steam generated from this turns the turbine to generate electricity. To prevent overheating of the reactor, a moderator is used to control the rate of nuclear fission reaction and a coolant to control the temperature. However, if the reactor overheats, it explodes and nuclear material leaks into the environment.

On April 25, 1986, an experiment was being conducted at this power plant to see how long the turbine could continue to generate power due to inertia when nuclear fission in the reactor was interrupted. Unfortunately, during the experiment on the No. 4 reactor, the worker did not adhere to safety regulations regarding the reactor's automatic cooling system. As a result, at 1:23 a.m. the next day, the reactor exploded due to overheating, releasing about 8 tons of radioactive material into the atmosphere. The reactor containment wall, made of thousands of tons of iron and

[1] https://en.wikipedia.org/wiki/chernobyl_disaster

S. S. Park, *Eco-disasters*, https://doi.org/10.1007/978-981-96-8667-4_25

concrete, was reduced to ruins in just a few seconds, and the nuclear fuel inside the reactor burned like a nuclear volcano at thousands of degrees. More than a thousand times the amount of radioactive material released here than when the atomic bomb was dropped on Hiroshima, Japan in August 1945.

Radioactive materials continued to leak from the exploded reactor for 10 days. They can cause diseases such as cancer and leukemia, and pregnant women can miscarry or give birth to deformed babies. The radioactive materials leaked at the time of the Chernobyl accident traveled thousands of kilometers from the accident site on air currents, crossing the Polish border and covering all of Europe, including southern Finland, Norway, Sweden, the UK and Spain, and parts of North Africa. The movement of radioactive materials released from Chernobyl to Europe was due to the rotation of the Earth and wind direction.

As a result of this accident, 31 power plant workers died on the spot and 42 more died from radiation poisoning afterwards. The surrounding buildings and ecosystems were contaminated by the leaked radioactive materials, forcing about 135,000 residents within 30 km of the power plant to evacuate, and putting 30 million people in Ukraine and Russia, who used the water from nearby lakes as a water source, at risk.

At the time of the accident, to suppress the fire and prevent radiation leakage, about 5,000 tons of lead and boron were dropped on reactor 4, and then covered with about 10,000 tons of concrete. Even after the incident, reactors 1 and 3 continued to operate, but on the night of April 24, 1996, another accident occurred in reactor 3. Radiation leaked due to a technician's mistake while replacing the exhaust filter used for air discharge from the reactor.

According to data observed from 1986 to 1990, the incidence of thyroid disease, cancer, leukemia, etc. in the Chernobyl radiation-affected area increased by more than 50%, and the rate of miscarriage, stillbirth, and birth defects also significantly increased. The property damage caused by the Chernobyl accident is estimated at 15 billion dollars, 90% of which occurred in the former Soviet Union, and the remaining 10% in neighboring countries. In the case of the former Soviet Union, Ukraine, Russia, and Belarus, which are adjacent to the accident site, suffered the most damage. In these three countries alone, about 9 million people were directly or indirectly affected, and at least 800,000 of them were classified as exposure risk groups. According to a 1995 UN report, various forms of damage, such as the birth of foals with two heads and babies without arms and legs, continued to appear here.

Among European countries, the damage was significant in southern Germany, Greece, Scandinavian countries, and the UK. The leaked radioactive materials continued to move on air currents, contaminating vegetables, fruits, and dairy products in neighboring countries, all of which had to be disposed of. At that time, Eastern European countries such as Poland, Romania, and East Germany, which were allies of the Soviet Union, issued bans on agricultural production, and anti-Soviet sentiment was It emerged. In countries like Denmark and Norway, which did not have nuclear power plants, it became a factor triggering anti-nuclear movements. And Germany was pushing for the construction of the Bakersdorf nuclear power plant at the time, but strong opposition began to form due to this incident.

The concrete sarcophagus that covered the accident reactor Unit 4 had a lifespan of only about 20–30 years, and the need for a new barrier began to be raised from the 2000s. Especially in 2010, the Ukrainian government, which confirmed the site where radioactive materials were leaking from the accident reactor and contaminating the nearby soil, began to design a new safety containment (NSC: New Safe Confinement) with a loan (1.5 billion euros) from the European Bank for Reconstruction and Development (EBRD) in 2012.[2] The installation was completed in 2016 by assembling two steel containments from the outside and pushing them over from next to the accident reactor. A total of 2.1 billion euros was invested, and the accident reactor is now completely isolated from the outside with a huge containment.

Today, most of the electricity we use comes from fossil fuels and nuclear power generation. Nuclear power generation is very environmentally friendly compared to fossil fuels if there are no accidents, but it leads to a large-scale environmental disaster in case of an accident. The Chernobyl accident, caused by the minor mistake of non-compliance with safety rules by workers, not only caused enormous damage to surrounding countries, but also raised questions once again about the safety of nuclear power, known as the great invention of mankind.

[2] https://www.ebrd.com/what-we-do/sectors/nuclear-safety/chernobyl-new-safe-confinement.html

Chapter 26
Bhopal Gas Catastrophe

Union Carbide, a multinational corporation from the United States, established a local factory in Bhopal, India, where it manufactured and sold pesticides. This factory stored a toxic gas called Methylisocyanate (MIC), a raw material for pesticide production, in underground tanks. MIC, used as a raw material for insecticides, herbicides, and pharmaceutical synthesis, reacts violently with water and even in small amounts can cause serious damage to human lungs and eyes, and instantly destroy the central nervous system and immune system. However, on December 3, 1984, around 12:30 am, an accident occurred where this toxic gas leaked out.[1]

While 750,000 Bhopal citizens were asleep, the toxic gas leaked out, amounting to a whopping 80,000 pounds (equivalent to 36 tons) over two hours, spreading to nearby residential areas. This gas, being heavier than air, settled like fog on the ground, and all living creatures within a 65 km^2 radius could not escape death. On the day of the accident, about 2,800 citizens died, and 17,000 suffered serious injuries, with many others developing various cancers and diseases, resulting in over 200,000 victims in total. This accident caused an enormous number of casualties in a short time and greatly damaged the natural ecosystem.

The number of deaths and injuries continued to increase afterwards. According to unofficial estimates by the Indian government, the death toll was estimated to be around 10,000, with about 600,000 injured, and 583,000 filed lawsuits for compensation. Union Carbide paid $470 million in compensation to the victims. However, most survivors had to live their lives in pain due to blindness, respiratory disorders, and abnormalities in the central nervous system and immune system. Also, because MIC alters human genes, the number of cancer patients greatly increased, and many babies were born with deformities.

This toxic gas storage tank, which always needs to maintain high pressure and low temperature, requires sophisticated safety devices and strict adherence to safety rules by the manager, as it poses an explosion risk if the temperature rises. However,

[1] https://en.wikipedia.org/wiki/bhopal_disaster

S. S. Park, *Eco-disasters*, https://doi.org/10.1007/978-981-96-8667-4_26

the storage tank in Bhopal lacked safety devices, safety rules were not followed, and the early warning system did not work properly. Investigators reported that the Bhopal accident was nothing short of a predicted disaster.

What shocked the world even more was the fact that they had boldly placed a toxic gas storage tank, which posed a constant risk, right in the middle of a densely populated urban slum. Moreover, there were two large hospitals and a railway station in close proximity. The residents of this area had been living in a defenseless state, unaware of the danger. This absurd situation eventually led to a disaster that claimed countless victims in an instant.

The Bhopal accident was the most fatal environmental disaster of the 1980s. It was incomparable to the Chernobyl nuclear accident, which claimed 31 lives at the time. No country in the world builds a nuclear power plant in the middle of a city. Even the storage of low-level nuclear waste, which is very low in radiation and very safe, is categorically rejected. However, the manufacture, storage, and transportation of toxic chemicals are generously allowed, and safety management is incomparable to that of nuclear power plants.

Following the Bhopal accident, many countries around the world began to investigate toxic substance-related accidents and their potential occurrence in their own countries. According to one investigation report, a total of 6,928 hazardous material-related accidents occurred in the United States from 1980 to 1985, resulting in 138 deaths, 4,717 injuries, and the evacuation of over 200,000 people. Two-thirds of the accidents occurred in manufacturing plants and storage facilities, and the rest occurred during transportation. The causes were mostly mechanical defects such as broken storage tanks or valves, or worker negligence. Furthermore, it was revealed that there are over 12,000 chemical plants and about 400,000 hazardous material storage tanks nationwide in the United States. Including laundries, paper mills, photo developing shops, and auto repair shops that use chemicals harmful to the human body, there are millions of hazardous material handling companies.

Especially after the Bhopal accident, a survey of factories or storage facilities located in densely populated areas revealed that there are over 6,300 related facilities in just 25 major cities where about 75% of the U.S. population lives. Among them, over 800 are located in the eastern U.S. megalopolis region from New York to Washington D.C. The New York Times reported on the status of hazardous materials in the United States after the accident, stating, "The fact that a major chemical accident causing numerous casualties has not occurred in this country is a great fortune. Instead of just trying to minimize the risk, we must change the system completely within a few years. Otherwise, there will be a huge disaster here as well." Although it may be an exaggerated expression, it is undoubtedly a vivid depiction of reality.

The Bhopal accident also brought changes to U.S. environmental law. In 1986, the U.S. Congress passed the 'Emergency Planning and Community Right-to-Know Act' was enacted. The reason for the enactment of this law was that the right of residents to know could be the most powerful means to prevent future disasters caused by toxic chemicals. Before this law was enacted, factories producing toxic chemicals or companies using them had no obligation to disclose to local residents what chemicals they use, where and how much they use. However, this law requires them to disclose the types and quantities of chemicals used and how they can be released into the environment in daily processes and in the event of an accident.

The purpose of this law is to ensure that industries using chemicals adhere to strict safety rules by disclosing all relevant information to residents. Since the enactment of this law, many companies have taken more precautions in managing chemicals and responding to accidents, and residents have also actively monitored companies for their lives and the environment. This law has not only brought very positive effects in the United States, but also greatly contributed to the enactment of similar laws in other countries.

So far, many developing countries have pursued policies to attract polluting industries operated by multinational corporations in developed countries to gain economic benefits. And this policy has spread rapidly in line with the interests of both developed and developing countries. However, in developing countries with insufficient disaster management capabilities, there is always a risk of accidents like Bhopal.

Today, mankind manufactures and uses about 100,000 kinds of chemicals, and it is known that about 2,000 new chemicals are developed every year. These chemicals are used everywhere from industry and agriculture to everyday life in homes and offices, and are so common that there is not a day without contact. However, about half of the chemicals we use are very harmful to all living things, including humans, and always pose a risk. From manufacturing and storage to transportation, use, and disposal, the safety management of hazardous substances is not an exaggeration to say that it is the lifeline of modern humanity.

Chapter 27
Rhine River Basel Chemical Spill

The Rhine River, originating from the Alps in Switzerland and flowing through France and Germany to the North Sea via the coast of the Netherlands, was the center of modern European civilization and a place that determined the fate of the world in the early twentieth century. Western civilization, which started in Greece and Rome, shifted its central axis to the Rhine basin after the Renaissance, and the Industrial Revolution, which started from the Thames in England, also bloomed with the Rhine as its base in the early twentieth century. This place was also the center of the two World Wars.

The Rhine River, with a length of 1,233 km and a total watershed area of only 200,000 km^2, is home to about 60 million people and is one of the most active industrial activities and high population density in Europe. Parts of seven countries including Switzerland, France, Germany, the Netherlands, Luxembourg, Belgium, and Austria, are located in the Rhine basin. Along the riverbanks, there is a high concentration of industries, including 10–20% of OECD member countries' chemical plants, as well as numerous steel and metal processing industries. Not only that, but intensive agriculture is carried out here, so there is a lot of use of pesticides and fertilizers compared to other basins, and 86 weirs with locks are installed along the river, and dozens of cargo ships and tourist ships operate from the North Sea to Switzerland or the Black Sea every day.

The present Rhine River is close to an artificial waterway that has undergone extensive civil engineering work several times. Originally, the upper reaches of the Rhine featured rugged valleys, while the lower reaches were characterized by fertile and expansive floodplains. There were meandering streams and more than 2,000 islands that repeatedly appeared and disappeared with winding tributaries. In the 1800s, a German civil engineer named Johan Tulla made an attempt to turn the pristine Rhine into an artificial waterway.[1] He changed numerous streams and islands into a single deep waterway, saying, "In principle, one river bed is enough for any

[1] When the Rivers Run Dry: Water–The Defining Crisis of the Twenty-first Century, Fred Pearce, Beacon Press, 2007.

S. S. Park, *Eco-disasters*, https://doi.org/10.1007/978-981-96-8667-4_27

river." As the Rhine turned into an artificial waterway, numerous industrial cities were built along the river to become the center of European industrialization. Currently the Rhine is one of the most actively utilized rivers in the world as a source of water supply, wastewater disposal, and a means of transportation.

However, the Rhine River has long suffered from various water pollution. Since the early twentieth century, warmongers advocating for national prosperity and strength through industrialization have operated in this area, leading to even greater devastation of the Rhine River. In fact, as soon as Hitler became the Chancellor of Germany in 1932, he encouraged national revival among the populace, proclaiming, "Let's build more factories and work hard until that blue Rhine River turns black." Just this fact alone can guess the situation at the time. By the end of World War II, there were frequent cases of fish dying en masse in the Rhine River due to various industrial waste and domestic sewage, and salmon, which are sensitive to water pollution, had already disappeared long ago.

As the pollution of the Rhine River reached a serious state, the countries located in the basin established the 'International Commission for the protection of the Rhine against pollution' in 1963 and attempted international cooperation for water quality management.[2] Centered on this committee, the 'Rhine River Chemical Pollution Prevention Treaty' and the 'Rhine River Chloride Pollution Prevention Treaty' were concluded in 1976. In addition, the committee installed water quality measurement stations in 49 places in the Rhine basin, including Switzerland, France, Germany, and the Netherlands. The pollution status was observed. The monitoring station continuously recorded flow rate, dissolved oxygen, pH, electrical conductivity, etc., and periodically measured organic matter, nutrients, heavy metals, radioactive materials, etc. In addition, a state-of-the-art biological warning system was introduced to immediately detect the influx of toxins or heavy metals in the event of an accident, prevent damage to the ecosystem, and stop the use of drinking water. This technology issues an alert by detecting the movements of fish, water fleas, luminescent microorganisms, shellfish, etc., that react sensitively when toxins are discharged into the river.

However, an absurd incident occurred on November 1, 1986, in the Rhine River, which had been operating a systematic water quality management and state-of-the-art warning system through international agreements.[3] A fire broke out in a warehouse located in Basel, Switzerland, at the uppermost part of the river, turning the Rhine River, the lifeline of 60 million people and the center of European industry, into a river of death overnight. The place where the fire broke out was a raw material storage warehouse of Sandoz, a multinational company in Switzerland that manufactures and sells pharmaceuticals, chemicals, pesticides, etc. It contained about 1,300 tons of 90 types of chemicals, most of which were highly toxic pesticides, sterilizers, fuels, and even 1.9 tons of toxic heavy metal mercury.

In Basel, there were several types of chemical manufacturing plants in addition to Sandoz, so special firefighting operations were needed in case of a fire, and special firefighters who received education and training for this were on duty. However,

[2] https://en.wikipedia.org/wiki/Management_of_the_Rhine_Basin.

[3] https://en.wikipedia.org/wiki/Sandoz_chemical_spill.

coincidentally, when the accident occurred, all the special firefighters were off duty, so extinguishing the fire was carried out using a normal fire method.

In places where toxic chemicals are present, sand or special extinguishing agents should be used to prevent secondary damage during fire suppression. However, because the general firefighters who did not know how to handle toxic substances used a large amount of water for extinguishing, the chemicals stored in the warehouse immediately flowed into the Rhine River with the water and seeped into the nearby soil and groundwater. Also, the toxic smoke generated by the fire caused great damage to the nearby residents and the natural ecosystem. As a result of this accident, 500,000 fish living in the Rhine River died en masse, and the benthic organisms completely disappeared in the 400 km downstream section of the accident site. Toxic chemicals stay long in the riverbed by being adsorbed to suspended matter, so the damage to the aquatic ecosystem lasted for a considerable period.

The Basel chemical spill, which can never be treated as a mere accident, has a lot to suggest to us. First, at the time of the incident, measures were in place to reduce the damage to the adjacent Rhine River when a fire broke out in the toxic chemical storage warehouse, but they could not be implemented due to lack of safety devices, and there were no second or third preventive measures prepared, so a single mistake led to an uncontrollable situation, and the damage spread widely. Fortunately, thanks to the state-of-the-art alarm system installed here, the leaked toxic substances were detected early and a quick water intake ban was issued to reduce damage downstream. In particular, the Rhine River mostly uses indirect water intake methods such as riverbank filtration, artificial recharge (charging river water into groundwater for intake), and drinking water reservoirs (treating river water and putting it back into the reservoir for reuse), so the damage to tap water production was minimal, and the weir that controls the water flow was continuously installed, which could slow down the downstream movement of pollutants.

As a result of this accident, Switzerland not only suffered enormous economic losses but also could not avoid criticism from neighboring countries. Sandoz, the cause of the incident, paid 100 million Swiss francs (equivalent to 60 billion won) in compensation to the victim countries such as Germany, France, and the Netherlands. After the accident, efforts to purify the Rhine River improved the water quality and ecosystem a lot, but the leaked toxic chemicals remained in the sediment for a long time. This accident reminded Europeans once again of the importance of the environment. Europe, which was engulfed in fear of radioactive materials due to the Chernobyl disaster in April of the same year, had to experience such an environmental disaster again in just a few months.

Chapter 28
Three Mile Island Nuclear Accident

In Harrisburg, the capital of the eastern state of Pennsylvania in the United States, flows a beautiful river named Susquehanna. This river passes through the U.S. capital, Washington D.C., and flows into the Atlantic Ocean. Near Harrisburg, on this river, there is a small island called Three Mile Island, where a nuclear power plant was built starting in 1968 and completed in 1974. At 4 a.m. on March 28, 1979, a nuclear reactor at the Three Mile Island nuclear power plant suffered a meltdown due to a cooling system failure, resulting in a leak of radioactive material.[1]

The accident originated from a failure in the cooling water supply system of the reactor. It was designed to supply cooling water through a secondary supply system in case of a problem with the primary supply system, but the valve of the secondary supply system did not operate. At the time, the control panel only displayed a signal that the valve was operating, but did not display information about the actual opening and closing of the valve, so the operator assumed that the valve was operating normally. However, in reality, the water supply was cut off, causing the reactor to overheat and a large amount of cooling water was supplied when the emergency core cooling system was activated. At the same time, nearly 100 warning lights on the control panel began to flash almost simultaneously. The confused operator misjudged the situation and made the mistake of reducing the amount of cooling water, causing the cooling system to rupture and the core to nearly half melt. This reactor had only been in trial operation for four months since December 1978.

As soon as the accident occurred, evacuation recommendations were first issued to pregnant women and children, and 23 nearby schools were closed. When the emergency evacuation order came, the residents trembled in fear. Fortunately, the accident was handled smoothly and there were no fatalities, but the reactor worth 2 billion dollars was destroyed in just 30 s. The infant mortality rate in the vicinity of Three Mile Island was more than twice as high as in other areas due to the radioactive material leaked at this time, and the Pennsylvania Department of Health announced

[1] https://en.wikipedia.org/wiki/Three_Mile_Island_accident.

S. S. Park, *Eco-disasters*, https://doi.org/10.1007/978-981-96-8667-4_28

that the number of newborns born with thyroid abnormalities, which are sensitive to radioactive materials, increased by more than 50%. The iodine-131 released during the accident is known to cause not only thyroid cancer in children but also intellectual impairment and growth retardation. In the accident area, a cancer incidence rate of 110 per 10,000 residents was also reported. This is recorded as the most serious nuclear power plant accident in twentieth century America.

In December 1942, the United States developed the world's first reactor, CP-1 (Chicago Pile No. 1), and began producing plutonium and enriched uranium. The United States used the radioactive materials produced in this reactor to manufacture nuclear weapons, and ended World War II by dropping them on Hiroshima and Nagasaki in Japan in August 1945. However, many lives were sacrificed due to this nuclear bombing, and the fear of humanity that started at this time still has a significant impact on the use of nuclear power. As if to atone for this horrific slaughter, the United States actively promoted the peaceful use of nuclear power after the war. In December 1953, President Eisenhower proposed the peaceful use of nuclear power at the UN General Assembly.

The world's first nuclear power plant was built in the Soviet Union in 1955, followed by the United States, the United Kingdom, and France. The United States built its first nuclear power plant in Shippingport, Pennsylvania in 1957. The reason for choosing Pennsylvania as the first nuclear power plant construction site was because it was the place where the Donora incident, in which about 20 people died from air pollution, occurred in October 1948. It meant replacing coal, which had caused past air pollution incidents, with new and clean energy.

The peaceful use of nuclear power pursued by the United States and the image of new and clean energy were very successful initially, and from the 1960s onwards, nuclear power generation continued to expand. The Three Mile Island power plant was built in this process. However, ironically, the worst nuclear power plant accident in American history occurred in Pennsylvania, where the Donora incident occurred and the first nuclear power plant in the United States was located. Before this accident, there was a fire at the Brown Ferry nuclear power plant in Alabama in 1975, but no radioactive materials were released and there were no casualties. However, the Three Mile Island incident, which was expected to cause casualties along with the leakage of radioactive materials, was a great shock to Americans who were immersed in the illusion that nuclear power is a safe and clean new energy. After this accident, the Three Mile Island power plant not only stopped operating, but the nuclear industry in the United States temporarily stagnated.

So far, the number of nuclear power plants has been increasing and the controversy has not ceased. The focus of the controversy is largely on accident risk, waste disposal, and issues related to the production of nuclear weapons using nuclear fuel. The waste from nuclear power plants is divided into low and intermediate-level waste and high-level nuclear waste. Low and intermediate-level nuclear waste is very low-risk waste, such as gloves, clothes, and shoes used by people working at the power plant. However, high-level nuclear waste, such as spent fuel and waste materials, control rods, is very dangerous if exposed to the human body. Typically, high-level nuclear waste must pass a period of ten half-lives (thousands of years). Also, high-level nuclear waste requires special surveillance because it can produce nuclear weapons if reprocessed.

Despite the technical safety of nuclear power generation and international surveillance, some countries are not easily accepting it due to fear and antipathy towards nuclear. In fact, in Germany and Sweden, the public sentiment opposing nuclear power was so strong that they even closed some nuclear power plants that still had a lifespan. However, most of the remaining countries continue to rely on nuclear power generation.

Now, humanity is in a very difficult position to choose the energy to use in the future. This is because neither nuclear power, fossil fuels, nor the third alternative are satisfactory. Fossil fuels are regulated due to carbon neutrality, and solar and wind power, often called eco-friendly energy, are not greatly welcomed alternatives considering their economic viability and space requirements. For example, to produce the same amount of power as a nuclear or thermal power plant through wind or solar panels, a vast amount of land is needed. It is calculated that several mountains with strong winds or intense sunlight would have to be cut down to obtain a similar amount of power. Therefore, some energy experts harshly criticize wind and solar power as "Dumb Energy". Moreover, wind power has problems with severe noise and bird protection, and solar power has harmful substances in the panels, causing another environmental problem when they need to be disposed of after their lifespan of 15~25 years. Bioenergy also requires a large amount of agricultural water and land, and securing these can lead to other environmental problems. Although many environmental problems have been improved, hydroelectric power is still not recognized as an eco-friendly energy.

The Three Mile Island accident was an environmental disaster without a single victim, but it completely changed the American fantasy that nuclear power is clean energy. And it made Americans, who were enjoying an abundance of energy in the 1970s, seek a new energy called Negawatts. Negawatts is a concept that the surplus energy created by efficient use and conservation of energy is new energy, also called the fifth fuel that humanity found after coal, oil, natural gas, and uranium. The world's largest energy underconsumption As the United States entered the 1980s, it began to develop energy-efficient homes, cars, and electronic products, and started to create a culture of energy-saving lifestyles.

Chapter 29
Seveso Toxic Cloud

On July 10, 1976, a large amount of toxic chemicals were released into the atmosphere due to overpressure inside a reactor at a chemical plant in a city called Seveso, located in northern Italy.[1] This chemical plant was a local Italian factory of Icmesa Chemical Co., a subsidiary of the Swiss pharmaceutical company Hoffmann-La Roche, and was producing a chemical called Trichlorophenol.

At that time, the toxic cloud created by the leak spread to 11 villages within 5 km of Seveso in just 15 min. Due to the chlorine gas contained in the leaked chemicals, 40,000 livestock died, and over 3,000 households were covered by the toxic cloud. Out of about 400 pregnant women, 51 miscarried due to this incident, and over 100 had abortions for fear of giving birth to deformed babies. Even the Roman Catholic Papacy, which prohibits abortion, issued a church decree allowing abortion for the victims of this incident. Many people were temporarily disfigured due to burns and skin diseases caused by the toxic substances. Children, in particular, suffered severe damage.

The leaked chemicals contained 2 kg of dioxin, a highly toxic substance, in addition to chlorine gas, contaminating the soil of the surrounding 18 km2. As soon as the incident occurred, an emergency evacuation order was issued and residents left the area. Even after the air cleared, access was prohibited due to the toxicity of the dioxin remaining in the soil. The government authorities, who could not anticipate the toxicity and scale of damage of the leaked chemicals at the time of the initial incident, issued an evacuation order for over 11,000 local residents only after 10 days, exacerbating the damage. As a result of this incident, the city center of Seveso was closed off, and it is expected that it will take 50 years for people to be able to live there again. The direct property damage from this incident is estimated to be about 250 million dollars.

The Seveso incident served as an opportunity to scientifically clarify the toxicity of dioxin. Dioxin began to be a problem as a defoliant in the Vietnam War in the

[1] https://en.wikipedia.org/wiki/Seveso_disaster.

S. S. Park, *Eco-disasters*, https://doi.org/10.1007/978-981-96-8667-4_29

late 1960s, but this was the first incident where it was leaked into a certain area at once and the damage could be scientifically investigated. Scientific research was conducted on the impact of the dioxin remaining in the soil of the Seveso region on the local wildlife and human body. Through this research, it was found that dioxin accumulates in the bodies of living organisms. It was revealed to be a substance that does not decompose and is highly concentrated, and is seven times more toxic than cyanide, which was known as the most toxic chemical until then. It was also revealed to be a substance that has a very high potential to cause cancer in humans, even in small amounts.

Dioxin is a substance that remains in the environment once it is emitted because it does not easily decompose in nature and does not oxidize even when burned at high temperatures. In this respect, dioxin has very similar chemical characteristics to PCBs and DDT, which have caused many environmental problems so far. The difference is that while PCBs and DDT caused problems by being intentionally produced and used, dioxin is produced as a byproduct of the production process or combustion of other chemicals.

Dioxin is strictly regulated today, but it is still being released into the environment through various routes, causing many environmental problems. It is produced as an impurity in the process of producing various organic chemicals, including herbicides, and is also included in the exhaust gas when burning garbage and is emitted together. In particular, dioxin is produced from chlorinated vinyl substances such as polyvinyl chloride (PVC) and food waste with a lot of salt. It is also known that dioxin is emitted from steel mills, paper mills, coal-fired power plants, and charcoal stoves.

The Seveso incident led to a political assassination later because the government, which should take swift post-disaster measures and fairly adjust compensation between perpetrators and victims, did not fulfill its role. Despite causing enormous damage at the time, the Icmesa chemical plant, the cause of the incident, did not try to compensate for the damage. The company's top executives publicly argued, citing the economic benefits they give to the region, that "capitalism means development and development can sometimes be inconvenient," angering the Italian people.

There were problems with the trial and compensation process of this incident. The court initially sentenced five officials responsible for the incident to prison terms ranging from two and a half to five years, but at the end of the trial conducted under the unstable political situation and corrupt government at the time, all but two were acquitted and the remaining two were released on bail after six months. And in December 1980, the compensation amount was determined to be $190 million, which is less than half of the actual damage. The government's lax post-disaster measures and the reality of corrupt politics in collusion with foreign capital revealed in this incident made all the people outraged. This gave an excuse for terrorism to the Italian left-wing forces that have been opposing capitalism.

One morning in February 1989, 13 years after the incident, Enrico Paoletti, who was the plant manager of the Icmesa chemical company in 1976 and the main person responsible for the Seveso incident, was shot and killed by a murder squad composed of three men and one woman. A few days later, an Italian left-wing terrorist group called 'Prima Linea' announced that they had killed him because they could not forgive him for causing enormous damage to the residents of Seveso and being released due to a wrong verdict.

Chapter 30
James River Kepone

In the 1960s, the American Allied Chemical Corp. developed a pesticide called Kepone, which was mainly sold overseas. Kepone is a highly toxic, non-degradable chemical with excellent adsorption used for cockroach extermination. This pesticide was exported to Central and South American countries to control rootworms in banana trees. In the United States, it was only used for cockroach and fire ant extermination, so the sales volume was not large, and with the development of more effective pesticides in the 1970s, the export volume of Kepone began to decrease rapidly. In 1974, Allied Chemical decided to stop producing Kepone, but two employees who worked at the company took over the sales network and patent and continued production.

They set up a company called Life Science in Hopewell, Virginia, and converted a closed gas station into a production plant. This company started producing Kepone from April 1974, but the working environment was very poor. Dust covered the manufacturing chemicals placed everywhere, and employees had to work for hours in the dust pit. After a few weeks, some employees began to show symptoms of poisoning, complaining of dizziness and neuralgia. Some of them showed severe symptoms requiring long-term hospitalization, and even their families showed similar symptoms. Eventually, more than half of the 150 employees suffered from headaches, visual impairment, liver disease, neuralgia, gastrointestinal disorders, convulsions, skin discoloration, chest pain, memory loss, and infertility. The Virginia Department of Health investigated and found that the toxic pesticide Kepone was the cause.[1] The United States Environmental Protection Agency (USEPA) then ordered the factory to be closed and all remaining Kepone sales to be stopped.

Ironically, the closure and sales ban led to the largest river pollution incident in the United States. When Life Science closed the factory in 1975, they dumped all the remaining toxic pesticides into the sewer. The pesticides dumped into the sewer flowed into the city's sewage treatment plant, killing the sewage decomposing

[1] https://virginiahumanities.org/2016/12/the-legacy-of-kepone/.

S. S. Park, *Eco-disasters*, https://doi.org/10.1007/978-981-96-8667-4_30

microorganisms and making sewage treatment impossible. The pesticide flowed into the nearby James River along with untreated sewage, which was even more disastrous. As a result, a 100 km stretch downstream of the James River turned into a river of death, devoid of any life forms.

The James River, a beautiful river flowing through the eastern state of Virginia in the United States and into the Chesapeake Bay of the Atlantic Ocean, was not only a source of water for local residents but also a major recreational area for fishing and boating. Moreover, the Chesapeake Bay on the Atlantic coast is a rich fishing ground, known as the most famous oyster production area in the eastern United States. However, due to this incident, the fish in the James River died en masse, and the oysters in the Chesapeake Bay became dangerous seafood. The Virginia state government ordered a fishing ban over a 160 km stretch of the James River basin and commanded not to consume or sell fish caught here. This ban was not lifted until 15 years after the incident, in 1989.

The estuary of the James River, where fresh water meets sea water, was a treasure trove of diverse species. In the case of estuaries, pollutants from the river cannot easily escape into the sea and stay for a long time, so if toxic substances that do not easily decompose are introduced, the damage to the ecosystem is significant. In the estuary, the flow of river water is blocked by seawater, and fresh water and seawater form layers due to density differences, preventing easy mixing. The heavier seawater with salinity and lower temperature sinks, while the fresh water coming down the river rises. Pollutants introduced into the estuary repeat sedimentation and floating between the two stagnant layers of fresh and sea water, causing damage to the resident organisms.

The U.S. Environmental Protection Agency made great efforts to purify the James River, but Kepone, being non-degradable and having strong adsorption to suspended matter, was detected in the riverbed sediments even after a considerable period, and the damage to the biota was not easily recovered. Especially, the damage in the estuary was severe and lasted for a long time. The research team that investigated the environmental damage at the time estimated that it would take decades for the estuary of the James River to fully recover.

A lawsuit claiming \$25 million in damages was filed over this incident, but Life Science, who should be responsible for it, was already bankrupt and unable to pay. Eventually, the responsibility for the incident fell back on Allied Chemical, and the company had to pay \$13 million in compensation to the Kepone victims and for the purification of the James River. Subsequently, in 1977, the Environmental Protection Agency banned the use of Kepone nationwide, and in 1978, it also banned its production and export abroad.

The contamination of rivers with toxic chemicals in the United States began in the 1950s. After the end of World War II, the use of synthetic organic chemicals increased, and their production also rapidly increased. As the use of cars and electronic products increased and products made from petroleum, such as plastics, became common, petrochemical factories were built all over the United States. Wastewater was often dumped into the rivers passing by these factories, and the environmental technology and systems of the time could not manage this.

The most stark manifestation of this problem occurred on June 22, 1969, in the Cuyahoga River incident. On this day, in Cleveland, Ohio, where the petrochemical industry was thriving, a massive fire broke out around noon on the river due to oil slicks floating on the surface, engulfing seven bridges in flames. This incident attracted national attention due to the skepticism of "how can a large fire occur in a river" and was received as a shocking event by Americans. The environmental movement, which was maturing in American society at the time, was greatly encouraged by this incident. Also, it played a decisive role in the establishment of the first Earth Day on April 22, 1970, in the United States. Moreover, this incident greatly contributed to the enactment of the Clean Air Act in 1970 and the Clean Water Act in 1972 by the U.S. Congress.

Incidents of rivers being contaminated with harmful substances are still the most frequently occurring environmental disasters around the world. It is commonplace for fish to die en masse due to toxic chemicals and for water production to be halted in downstream areas. This usually occurs due to complacent thinking that it will be purified at the sewage treatment plant or unscrupulous behavior that it won't be noticed if discarded. Although the system is being strengthened to prevent accidents and wastewater discharge is being more strictly monitored, accidents are constantly recurring.

The James River Kepone incident conveys several important messages to us. First, factories dealing with harmful substances must be closed according to strict regulations and procedures. While systems are in place to review the impact on the environment when establishing a factory, issues that can arise during closure or relocation are often neglected. Second, more stringent monitoring of lax urban sewage management is needed. There are limits to the substances and capacity that can be filtered out at sewage treatment plants, and exceeding these limits leads to water pollution accidents. The possibility of toxic substances entering the complex web of urban sewage networks and leading to accidents is very high. Third, proper environmental education is required for citizens. The environment Accidents can sometimes occur in unavoidable situations, but they are usually due to ignorance about the environment. Therefore, it is necessary to correctly educate citizens about the process through which the water they use and discard returns to nature. No matter how much surveillance activities are strengthened under a strict system, environmental accidents can occur at any time without the voluntary participation of citizens.

Chapter 31
Goiânia Cesium Tragedy

In 1985, a private hospital in Goiânia, Brazil, built a new building and moved. This hospital was a cancer treatment center that had been using an old cesium-137 radiation therapy machine in an old building in the city center, but when it moved to the new building, it purchased a new cobalt-60 therapy machine. The cesium-137 therapy machine was left in the old building because there was no proper place to dispose of it.[1] The hospital had to report to the government authorities about the purchase and disposal of the radiation therapy machine, but it only reported the newly purchased one and did not mention the old therapy machine. The Brazilian government authorities sent two inquiries by mail asking where and in what condition the previously used cesium-137 therapy machine was, but there was no response.

Two years after the radiation therapy machine was left in the broken old hospital, Roberto dos Santos Alves and Wagner Mota Pereira, two young men who had no regular jobs and were idling around, saw it. On September 13, 1987, they detached the heavy radiation therapy machine attached to the ownerless building and sold it to a junk dealer for 25 dollars. The junk dealer and his employees disassembled this therapy machine over several days, and around September 18, they broke the lead capsule and took out the contents. The material inside the capsule was none other than cesium-137, a small grain of about 20 g that glows blue in the dark. Cesium-137 has a half-life of about 30 years, which is relatively short compared to other radioactive materials, but it emits powerful radiation. This material is also found in nuclear waste, and people exposed to it for a long time lose their lives.

The junk dealer found it fascinating that the small grains inside the capsule glowed blue in the dark, so he distributed them to people around him. People who considered these as valuable gems wrapped them in handkerchiefs, put them in their trouser pockets, and also kept them at home for viewing. Also, because the hands or skin that touched these small grains also glowed, some people rubbed them on their skin for fun. Especially, the junk dealer's brother, who received a lot of these grains, gave

[1] https://en.wikipedia.org/wiki/GoiAnia_accident.

S. S. Park, *Eco-disasters*, https://doi.org/10.1007/978-981-96-8667-4_31

them to his wife and six-year-old daughter, and they rubbed the grains on their skin or ate sandwiches with the hands that touched them. These grains crumbled into powder, stuck to the skin, and went into the mouth.

About a week after the capsule was opened, many people began to show symptoms of digestive disorders, colds, and hair loss, and were also seized with vague anxiety. On the hands and skin, Wounds were formed and anyone who ingested even a small amount of the substance had severe erosion in their mouth, tongue, and esophagus. On September 28, the owner of the junk shop, who had severe symptoms, was the first to visit the local health center. He suspected the blue granules and took them in a bag on a bus to the health center. The health center examined the substance brought by the junk shop owner and the symptoms that appeared on his body, realized that a serious radioactive material leak had occurred, and reported this fact to the Brazilian Nuclear Energy Commission.

The Brazilian authorities formed an emergency response committee and asked the International Atomic Energy Agency for help with post-accident measures and patient treatment. They also began investigating the process of radioactive material leakage, victims, and the status of contaminated areas. The authorities gathered 112,000 residents, 10% of the population in the Goiânia area, in a soccer stadium and investigated contamination with a radiation meter. As a result, about 250 people were found to be contaminated with radiation, and about 50 people with severe degrees were isolated for detailed investigation. Among them, about 20 people were hospitalized and 8 were sent to a naval hospital in Rio de Janeiro. Among them, four people, including the wife and daughter of the junk shop owner's brother, all died within a month of admission. In addition, 28 people suffered serious skin damage from radiation and 14 showed abnormalities in bone marrow cells. A year or two later, three more people died of cancer, and the junk shop owner also died in 1994.

As a result of the investigation in the 67 km2 area of Goiânia, eight places were contaminated with radiation, and it was confirmed that radioactive materials had moved up to 160 km away from the accident site. This was because it rained between September 21 and 28, spreading the radioactive material widely. In particular, 85 households in the survey area were seriously contaminated, and more than 200 people from 41 households were urgently evacuated.

The purification work for the contaminated area and houses continued until the end of December 1987. The 85 houses that were heavily contaminated with radioactive materials were demolished and the soil was also removed. The problematic radioactive waste was all collected and isolated, with a volume of as much as 3,500 m3. Also, the bodies of people who died in this accident were placed in lead tubes that radiation could not penetrate, rather than being cremated or buried like other corpses, because the radioactive material in the body could leak into the environment. They were placed far away from residential areas.

In March 1988, a medical research institute was established in Goiânia to manage the damage caused by this incident. The government provided pensions to 111 victims who were heavily exposed to radiation. They could not come into contact with people or major products produced in Goiânia because radioactive substances could be emitted from their bodies. Therefore, the victims had to suffer from serious mental pain such as fear and depression. The Brazilian government spent about 20 million dollars on the aftermath of this incident. However, the damage suffered in agriculture, industry, and tourism was estimated to reach hundreds of millions of dollars. The price of products produced here plummeted by 40% for a month or two after the accident, and many residents left the place altogether.

Citizens staged large-scale protests, angry at the government's irresponsible management of radioactive materials. Citizens sympathized with the person who took the abandoned radiation therapy machine or the junk dealer, saying that the poor might have had no choice but to make a living, and argued that all responsibility should be returned to the hospital that neglected the therapy machine and the government. Eventually, the responsibility for the incident fell on the three doctors who jointly operated this hospital which had the duty to manage radioactive materials, along with the Brazilian Nuclear Energy Commission, the Ministry of Health, and the Ministry of Labor.

Goiânia cesium tragedy has been recorded as one of the most serious radioactive material accidents in the world to date, leaking a large amount into the natural environment after the Chernobyl disaster. The amount of radiation leaked from this accident reached a total of 1,400 curies. This is not comparable to the Chernobyl disaster, which emitted 50 million curies, but it is more than the amount of small-scale nuclear power plant accidents, including Three Mile Island. Especially in this accident, people directly touched radioactive materials, put them in their mouths, put them in bags, and used public transportation, which greatly amplified the damage compared to other accidents due to very severe exposure. Also, despite the fact that the victim's body itself was a source of radiation emission, the manager came into direct contact with the victim and used a regular ambulance due to carelessness in the initial response, causing secondary contamination. Therefore, several police officers, doctors, and nurses also became victims of radiation from this accident.

The Goiânia tragedy was made into a real-life movie titled "Cesium 137: The Nightmare of Goiânia" in Brazil in 1989, which runs for 1 h and 35 min. The movie script was based on the testimonies of victims and medical staff, and although it is a Portuguese movie used in Brazil, it has been screened worldwide for a long time as an environmental movie that informs about the dangers of radioactive materials with English subtitles. Especially, it was screened at the film festival, a side event of the UN Environment Summit (Rio + 20) held in Rio, Brazil in 2012, and was greatly appreciated. It attracted attention.

The Goiânia tragedy served as a wake-up call for the world about the everyday use of radioactive materials. Radioactive materials are widely used in numerous hospitals, research institutes, universities, and industries around the world, and their number is so enormous that it is difficult to count. In the United States alone, there are approximately 2 million places using radioactive materials, and it is estimated that a quarter of them are left unused and neglected. Therefore, if there is no knowledge about these materials and systematic management is not carried out, accidents can happen anytime, anywhere.

In particular, radioactive materials have a very high potential to be used in terrorism. In the conflict-ridden former Soviet Chechnya, there have already been several instances of terrorism using radioactive materials, and in 1998, a theft of radioactive materials for terrorist purposes occurred in a hospital in Greensboro, North Carolina, causing a stir across the United States. The world is on alert for small bombs using radioactive materials, commonly known as “dirty bombs”. This is a combination of conventional bombs with radioactive materials such as cobalt 60, strontium 90, cesium 137, iridium 192, radium 226, plutonium 238, etc., which can be easily manufactured and can contaminate with radioactivity up to several kilometers in radius.

Radioactive materials were first discovered by French scientists Antoine Becquerel, Marie Curie, and Pierre Curie at the end of the nineteenth century. Becquerel discovered radioactivity from uranium compounds in 1896, and Marie Curie, along with her husband Pierre, separated plutonium from uranium ore in 1898 and discovered radium, which emits intense radiation. For this achievement, they jointly won the Nobel Prize in Physics in 1903. Marie Curie’s research on radioactivity continued, and she won the Nobel Prize in Chemistry again in 1911. However, she died of leukemia due to exposure to high-level radioactivity during her research, and her daughter Irène Joliot-Curie also tragically died of leukemia.

Just as radioactive materials brought both the glory of the Nobel Prize and the tragedy of leukemia to the Curie family, they are giving both to our humanity. Nuclear power generation, which began in the late twentieth century, has become a new energy source replacing fossil fuels and is saving countless lives when used for treatment. However, Nuclear accidents, nuclear waste, the risk of terrorism, and nuclear weapons of some rogue states are threatening the existence of modern humanity. Radioactive materials, which have become a double-edged sword that modern civilization cannot avoid, are demanding careful management from all of us.

Chapter 32
Nakdong River Phenol Leak

In March 1991, an accident occurred in Gumi, upstream of the Nakdong River in South Korea, where the toxic chemical phenol was leaked.[1] The phenol leaked into the Nakdong River flowed downstream, and the first abnormal phenomenon was detected in Daegu, South Korea's third-largest city that uses this water as a tap water source. During the chlorine disinfection process at the water purification plant, the phenol changed to chlorophenol, causing severe abnormalities in the taste and smell of tap water. Eventually, citizens who drank tap water complained of strange tastes and smells, and as a result of tracing the cause, it was revealed that there had been a phenol leak accident at an upstream industrial city. The tongues and noses of the citizens played the role of a toxic chemical leak detector. The phenol was already leaked on March 14, and the citizens complained about the abnormality of tap water two days later, on March 16. The place where the accident occurred was D-Electronics, located in the Gumi Industrial Complex.

At that time, D-Electronics used a spare underground pipe instead when the usual above-ground pipe used to supply phenol raw material from the storage tank to the production process broke down. However, about 30 tons of phenol raw material leaked from the pipe connection, which flowed into the Nakdong River through the drain. The phenol that contaminated the tap water in Daegu City did not stop there, and flowed downstream driving the residents of the entire downstream region, including Busan, the second largest port city, into phenolic fear.

Phenol is a toxic chemical that causes various cancers and abnormalities in the nervous system, and along with PCBs and heavy metals, it has often been a problem in industrial wastewater. Also, this substance easily detected through tap water because it turns into a substance called chlorophenol, which causes a severe odor during the chlorine disinfection process of tap water. However, the place where this substance was leaked did not care about the damage to the downstream residents and was eager

[1] https://www.nytimes.com/1991/04/16/business/chemical-leak-in-korea-brings-forth-A-new-era.html.

S. S. Park, *Eco-disasters*, https://doi.org/10.1007/978-981-96-8667-4_32

to hide it for fear of their responsibility. If immediate action had been taken as soon as the leak accident occurred, the damage could have been greatly reduced.

Many citizens who drank contaminated tap water at the time of the phenol incident complained of damage, and a total of 1,617 cases of damage were reported in Daegu alone, and especially, the furies of pregnant women who were afraid of giving birth to deformed babies were unbearable. The government ordered D-Electronics to suspend operations and improve facilities, especially the underground supply line of phenol raw material. The pipes were closed and the above-ground pipes were repaired and reinforced. The water supply, which had been suspended, was resumed on March 23.

However, the government suddenly decided to resume operations on April 9, and D-Electronics restarted on April 18. This led to the second phenol leak accident. During normal operation after a test run, about 1.3 tons of phenol raw material leaked again on April 22 due to a failure in the above-ground pipe joint. This time, about 1 ton of phenol raw material was recovered through initial response, greatly reducing the damage. However, the citizens' anger reached its peak due to the consecutive leak accidents. Due to the second phenol leak accident, D-Electronics was ordered to suspend operations for 64 days (from April 27 to July 1). They were also ordered to install a discharge wall to prevent leaks and to enhance remote monitoring and automatic alarm devices for leak accidents. The water supply was temporarily suspended again, but resumed on April 24.

As a result of this accident, D-Electronics compensated about 22 billion won, and the Minister and Vice Minister of the Environment, who hastily decided to resume operations, were forced to resign. In addition, seven public officials who tried to cover up the incident were arrested. The public's anger led to a nationwide boycott of D-Group products. In particular, citizens of Daegu launched a movement to condemn the government for poor management of the Nakdong River water quality and supplying polluted tap water, and to refuse to pay water bills. D-Group spared no investment in environmental facilities and environmental education at the group level to shake off the stigma of being a phenol company immediately after the incident, but the damage caused by the product boycott was enormous.

This accident is known as the environmental disaster in South Korea that has incited the most intense public outrage and had the greatest social impact to date. This is because it contaminated the tap water, which is vital to citizens' lives, with carcinogens, affecting an unspecified large number of people. The immorality of the large corporation involved exacerbated the damage. Moreover, this incident instilled the awareness that water is directly linked to the right to survival of the people, and it prompted the government to actively implement water quality policies.

Following this incident, the government implemented strong water management measures, starting with the 'Clean Water Supply Comprehensive Plan' in 1993. Major environmental infrastructure projects such as sewage treatment plant construction, sewage pipe maintenance, and polluted river purification were promoted. Advanced water management systems and technologies, such as total pollution load management, designation of riparian zones, land purchase in water source protection areas, installation of automatic measurement networks, and integrated watershed management systems, were introduced.

Incidents similar to the Nakdong River phenol leak have been repeated numerous times, mainly in advanced industrial countries. Incidents where harmful chemicals flow into rivers, causing mass fish deaths and contaminating water sources, leading to the suspension of tap water supply, were a common menu of environmental disasters. Advanced industrial countries have been reducing damage by developing various technologies and policies, from accident prevention to post-accident monitoring, as incidents are repeated.

However, water pollution accidents are still ongoing in developing countries. The Songhua River benzene leak in 2005 is one of the largest river pollution incidents in China to date.[2] On November 13, a series of explosions and fires occurred at the benzene factory of Jilin Petrochemical Company in Jilin Province, China, causing significant casualties on the spot (6 deaths, 6 missing, 121 injured) and evacuating 1.25 million residents. This accident caused about 100 tons of benzene to flow into the Songhua River, creating a long pollution belt of 80 km, and water production was suspended in downstream Harbin, China, and Khabarovsk, Russia. China suffered enormous damage during the rapid urbanization and industrialization process because it was careless with environmental management and had no measures to detect and prevent the spread of accidents early when they occurred. In July 2010, another accident occurred where about 7,000 drums containing flammable liquid from another chemical factory in Jilin City were discharged into the Songhua River due to a flood.

[2] https://en.wikipedia.org/wiki/2005_Jilin_chemical_plant_explosions.

Chapter 33
Tokaimura Fukushima Nuclear Accidents

Tokaimura in Ibaraki Prefecture, Japan, was a small fishing village located on the northwest Pacific coast of Tokyo. In 1956, Japan built a nuclear research institute here, and since then, nuclear fuel development teams and power plants and other related facilities have been established. This small fishing village later transformed into a center for nuclear research in Japan, which continues today.

On September 30, 1999, a radioactive leak occurred at the Tokaimura uranium processing plant.[1] The accident was revealed to have occurred when a private company, JCO (Japan Nuclear Fuel Conversion Co) Tokai, injected more uranium than allowed into a precipitation tank used to dissolve uranium impurities with nitric acid, causing a chain nuclear fission. They violated the rule of injecting only up to 2.4 kg at a time and injected about 7 times more, 16 kg of uranium. If the chain nuclear fission caused by this uranium reaches its peak, it could lead to a massive explosion like a nuclear bomb.

As a result of this accident, an evacuation order was issued for 310,000 residents living within 10 km of the accident site, and schools and public offices were closed and traffic was controlled in five nearby cities. Immediately after the accident, the Japanese government stopped all activities and focused on establishing accident countermeasures, and all broadcasters stopped regular broadcasting and reported the accident news all night. Fortunately, the recovery work went smoothly, and the situation was resolved 20 h after the accident, but at this time, 49 people were directly exposed to radiation. Among them, three were directly exposed to radiation at the accident site and fell into a coma, and one of them, Hisashi Ouchi, died at the age of 35 at Tokyo University Hospital where he had been receiving treatment 83 days after the accident, on the night of December 21.

Since starting nuclear power generation in 1957, Japan has experienced several radioactive leaks. In particular, Tokaimura has experienced more than 20 radioactive leaks since the first accident in 1971, causing over 100 victims. For example, in

[1] https://en.wikipedia.org/wiki/Tokaimura_nuclear_accidents.

S. S. Park, *Eco-disasters*, https://doi.org/10.1007/978-981-96-8667-4_33

March 1997, a fire at a reprocessing plant caused a radioactive leak, exposing 37 people, and in August 1997, more than 2,000 drum cans for storing spent fuel in underground storage facilities corroded, causing a nuclear material leak. In addition, in Japan, there was a Mihama nuclear power plant accident of Kansai Electric Power in February 1991, and in December 1995, volatile liquid sodium. Accidents involving the leakage of one ton of nuclear material have been occurring continuously in Japan. The Power Reactor and Nuclear Fuel Development Corporation (DONEN), the central institution of Japan's nuclear development with a 30-year history, was forced to shut down in 1998 due to repeated accidents.

Japan, more than any other country in the world, is well aware of the horrors of nuclear power due to the atomic bomb damage in Hiroshima and Nagasaki in 1945. However, Japan has built many nuclear power plants to date because there was no other way to supply the massive energy needed by the country without using nuclear power.

The Tokaimura accident was a huge shock to the Japanese people. The reason was not because the scale of the damage was the worst in history, but because it showed that despite repeated nuclear power plant accidents since 1971, the government's measures and the safety consciousness of the managers had not improved at all. Moreover, it was a great tragedy for Japan, the only country in the world that has suffered from atomic bombs, to end the twentieth century with a nuclear power plant accident.

This was not the end of Japan's nuclear power tragedy. On March 11, 2011, at 14:46, the most powerful earthquake in Japanese history, with a magnitude of 9.0, occurred in the eastern seabed of Miyagi Prefecture in northeastern Japan (38.1 degrees north latitude, 142.9 degrees east longitude, and a seabed depth of 23.7 km). At the time of the earthquake, there were a total of 54 commercial nuclear power plants in Japan. Of these, the nuclear power plants directly affected by the earthquake were, in order of proximity to the epicenter, the Onagawa nuclear power plant (3 units) in Miyagi Prefecture, the Fukushima Daiichi (6 units) and Fukushima Daini (4 units) in Fukushima Prefecture, and the Tokai No. 2 nuclear power plant (1 unit) in Ibaraki Prefecture, a total of 14 units. Among them, units 4, 5, and 6 of the Fukushima Daiichi nuclear power plant were shut down for nuclear fuel exchange and regular inspections, and the remaining 11 units that were in normal operation were automatically shut down immediately after the earthquake. Other nuclear power plants in Japan continued to operate without being affected by the earthquake.

About 50 min after the earthquake, a 15 m high tsunami hit the Fukushima Daiichi nuclear power plant. This was far beyond the maximum tsunami height of 5 m assumed in the power plant design. The emergency diesel generators installed in the basement were flooded and stopped by the incoming tsunami, and all the electrical facilities of the power plant were damaged. The Fukushima Daiichi nuclear power plant fell into a blackout state without even the minimum power for reactor safety, and as a result, it became impossible to operate the cooling water pump for reactor cooling.

When the power supply to the cooling system was interrupted, the cooling water rapidly evaporated, causing the reactor The internal temperature and pressure rose, eventually causing the reactor to explode and a fire to break out in the pool where spent fuel rods were cooled, resulting in a large amount of radioactive material leaking out in gaseous form. Not only that, the seawater that was sprayed to cool the heated reactor in place of the halted cooling system contaminated the groundwater with radioactive materials and flowed into the sea.

As a result of the accident, the Fukushima area and nearby coasts turned into serious radioactive contamination zones. This incident was Japan's worst nuclear accident, comparable to the explosion of the Chernobyl nuclear power plant in the former Soviet Union. The shock of the incident had a significant impact on many country's nuclear phase-out policy, which had been questioning the safety of nuclear power generation. Especially, the aftermath was greatly amplified as nearly 20,000 deaths and missing persons from the tsunami at the time, and hundreds of thousands of displaced persons, were mistakenly thought to be caused by the Fukushima nuclear accident.

A notable fact is that there were no casualties (0) due to exposure from the Fukushima nuclear accident. The UNSCEAR (United Nations Scientific Committee on the Effects of Atomic Radiation) report 2013, released two years after the accident, also concluded that no one died from radiation. The reactor automatically shut down immediately after the earthquake, and about 50 min later, the tsunami hit and the reactor exploded, so no one was there. Also, because all the residents had evacuated from the radioactive contamination area, they were able to prevent damage. According to the investigation reports released so far, the cancer incidence rate in the Fukushima area does not show a significant difference compared to other areas in Japan.

Part VI
Dangerous Modern Living Environment

Chapter 34
Philadelphia Legionnaires Outbreak

Philadelphia, located in the eastern part of Pennsylvania in the United States, is a city of great historical significance. The Declaration of Independence was made here in 1776, and it was the capital of the federal government from 1790 to 1800. The 13 states in the eastern region that declared independence from Britain formed a federation in 1781, and with the inauguration of the war hero George Washington as the first president in 1789, the United States finally established a national system.

The inauguration and work of the first president began in New York, but in accordance with the agreement that the capital should be where the Declaration of Independence was made, Philadelphia became the capital of the United States in 1790. However, President George Washington judged that it would be desirable to build a new city for the future of the federal government, and from 1790, when he started working in Philadelphia, he planned a new city and prepared for relocation. The location of the new city was determined to be the Potomac Riverside which is geographical in the middle of the 13 states and facilitates westward expansion along the waterways. The capital relocation was carried out by the second president, John Adams, in 1800, and the capital was named Washington D.C. (Washington District of Columbia), after the first president who died in 1799 and Christopher Columbus, who discovered the American continent.

Philadelphia was the capital for about 10 years, but historical artifacts symbolizing American independence, including the Liberty Bell, are still preserved, and many events are held every year on Independence Day. Many events were held in Philadelphia around July 4, 1976, the 200th anniversary of American independence, one of which was a veterans' meeting held at the Bellevue-Stratford, a luxury hotel, from July 21 to 24. However, among the 4,400 attendees of this meeting, 255 people showed severe pneumonia symptoms with high fever and cough between July 22 and August 3, and 34 of them died.[1]

---

[1] https://en.wikipedia.org/wiki/1976_Philadelphia_Legionnaires%27_disease_outbreak.

S. S. Park, *Eco-disasters*, https://doi.org/10.1007/978-981-96-8667-4_34

The U.S. Centers for Disease Control and Prevention (CDC) entered an epidemiological investigation to identify the cause of the disease. Although all the patients showed malignant pneumonia symptoms, the investigation revealed that it was not pneumonia. Therefore, Initially, it was reported in the media as a mystery disease. The Centers for Disease Control and Prevention announced that the cause of the disease was a new bacteria that had not been known before, which had contaminated the indoor air of the hotel. And they named this new bacteria (pathogen) in relation to this incident as Legionella pneumophila. The name of the pathogen includes the meaning of the Philadelphia Veterans Association where the incident occurred, with Legion (military unit), ella (bacteria), pneumo (lung), and phila (Philadelphia). Legionella bacteria typically inhabit rivers, lakes, reservoirs, etc. at very low concentrations, but the number of individuals suddenly increases when the temperature is right. At the time of the incident, Legionella bacteria massively proliferated in the air conditioning cooling water and contaminated the indoor air of the hotel through the ventilation system. This incident caused a big social ripple, enough to be featured as a cover story in famous American weekly magazines such as "Newsweek" and "Time". The reason is that the medical clarification of the mystery disease was first made while the world's attention was focused on Philadelphia for the 200th anniversary of American independence.

Legionnaires' disease, which was revealed in this way, has since been reported as a disease that frequently occurs in various parts of the world, including France, the UK, and Spain. Before the incident, it was mistaken for flu, pneumonia, or passed as an unknown disease. Since then, it has continued to increase, and now it is estimated that 8,000 to 18,000 people in the United States contract this disease each year, and about 12% of them die. The cause is the inhalation of Legionella bacteria through air conditioning wind, or steam from tap water containing Legionella bacteria during bathing or showering. Health authorities classify it as a disease that requires special attention because there is no vaccine developed yet, although human-to-human transmission does not occur. Interestingly, about 70% contracted the disease while traveling, and some occur in hospitalized patients. This phenomenon shows that healthy people have resistance to Legionella bacteria, but they can get sick if they are tired or their immunity drops. Also, the incidence is high in people who smoke or have unhealthy lungs.

In June 2016, the American daily newspaper Washington Post reported data from the Centers for Disease Control and Prevention that the incidence rate had increased fourfold from 0.42 per 100,000 population to 1.62 between 2000 and 2014 over 15 years.[2] The reasons for this sharp increase since 2000 include an increase in the elderly population and travelers, an increase in poor water pipes due to aging buildings, and an increase in testing and diagnosis. Here The rise in temperature due to urbanization also provides good conditions for the growth of Legionella bacteria in water pipes and tanks, which is one of the reasons, they said.

[2] https://www.washingtonpost.com/news/to-your-health/wp/2016/06/07/legionnaires-outbreaks-cases-nearly-quadrupled-in-15-years/ .

This incident suggests various important environmental issues that modern people can face, such as respiratory diseases from tap water, the harmfulness of indoor air, and buildings that kill people. Especially in modern society, we are in contact with tap water every day and spend most of the day indoors, so if conditions for the growth of Legionella bacteria are provided, the risk of disease is always lurking. For prevention, careful management of tap water in hospitals, hotels, motels, nursing homes, and bathhouses, and maintaining clean indoor air is necessary. Since the optimal breeding temperature of Legionella bacteria is 25 ~ 42°C, managing air conditioning cooling towers and HVAC (Heating, Ventilation, & Air Conditioning) systems, especially in the summer, is most important.

If indoor air is poorly managed, it can directly cause respiratory damage to the human body. A report from the World Health Organization estimates that the annual number of deaths due to air pollution can reach up to 6 million, and about 2.8 million of these are estimated to be caused by indoor air. It is particularly pointed out as one of the main causes of infant mortality in developing countries due to indoor air pollution. Also, because the probability of pollutants released indoors being delivered to people's lungs is 1,000 times higher than outdoors, reducing indoor pollution by only 20% can reduce the mortality rate of acute bronchial diseases by at least 4 ~ 8%.

Common major sources of indoor air pollution include air conditioners, household goods, and building materials. Bacteria living in air conditioners and carbon monoxide, nitrogen oxides, and sulfur dioxide produced by heating facilities also threaten health. Indoor air pollutants emitted from various detergents, oil removers, synthetic fibers, and household insecticides are also fatal to the human body. Especially, the damage from air pollution caused by cooking in the kitchen is surprisingly serious.

This outbreak remains as a case that informed the world that high-end hotels can make many people sick and lead to death. The Bellevue-Stratford Hotel, where the incident occurred, subsequently closed its doors. It reopened in 1979 under the name The Fairmont Philadelphia, but could not operate normally. After changing the hotel name and closing and reopening several times, it has been operating under the name Hyatt at The Bellevue since 2010. This hotel, built as a 19-story building in 1904, was a remarkable building at the time, and many major events that have been recorded in American history over a long period of time were held here, and it is currently designated as a national historical artifact. However, this incident left a lesson that no matter how great a building is, if environmental management is poor, it can turn into a haunted house where people get sick and die, leading to its downfall.

Chapter 35
Stanley Watras Radon Story

The Limerick nuclear power plant, located on the Schuylkill River, a tributary of the Delaware River in Pennsylvania, USA, was about to start operations on December 22, 1984, after a long period of construction. Three weeks before operation, a radiation detector was installed at the entrance where the power plant employees commute. This is a basic facility of all nuclear power plants, a device to prevent employees from accidentally taking radioactive materials out of the power plant on their clothes.

On December 2, when the detector was just installed, an alarm sounded as construction engineer Stanley J. Watras passed through the radiation detector on his way to work. At that time, the nuclear power plant was not yet operational, so no nuclear fuel had been installed. Moreover, it was incomprehensible that an alarm would sound on the way to work, not on the way home. The new detector was obviously fine, and there was no alarm when other employees passed through. Watras's alarm on his way to work continued every morning for two weeks. This mystery was finally solved when the radiation safety officers of the nuclear power plant visited Watras's home directly and found the radioactive material. The cause was that Watras had gone to work with radioactive material on his clothes from home.[1]

The radiation safety officers made a very surprising discovery while searching for the source of contamination in Watras's house. An enormous concentration of radon gas was measured in the basement. About 2,700 picocuries (pCi/l = 99,900 Bq/m3 becquerel) of radon were detected. Radon (Rn-222) is a naturally radioactive material that is released into the air from granite, soil, and groundwater. It is a class 1 carcinogen that causes lung cancer in a state that humans cannot perceive at all because it is invisible and has no particular smell or taste. The radon in the basement of Watras's house was 20 times higher than the concentration usually observed in uranium mines. The harmfulness of radon there was estimated to be the same as the possibility of developing lung cancer when smoking 135 packs of cigarettes a day.

[1] https://radon-ohio.com/the-stanley-watras-story/.

S. S. Park, *Eco-disasters*, https://doi.org/10.1007/978-981-96-8667-4_35

As soon as this fact was revealed, the Watras family moved out immediately. They had lived here for almost a year since they moved to this house in January 1984. The U.S. Environmental Protection Agency (EPA) and the Pennsylvania state government turned this house into a radon and radon decay product laboratory and studied unusual phenomena that could not be seen before. And they studied whether people could live in this house again without health damage. They also studied the radon reduction method (Mitigation Method) to create a safe living environment.

Radon is a substance that comes out when radium, which is created by the slow nuclear fission of uranium contained in underground bedrock, undergoes nuclear fission again. The radon thus produced is a gaseous substance that has been released into the basement. Interestingly, high concentrations of radon gas were detected in the basement of the Watras house, but the neighboring houses did not exceed the safety standard. The investigation team concluded that a radon gas vein was formed in the bedrock there, like an underground water vein, and was being released into the basement of the Watras house. In the basement of that house, radon gas was gushing out like spring water.

The radon reduction research team repaired the Watras house for several months and made it a safe place for people to live. By connecting the underground radon gas outlet to a pipe and blowing it into the outdoor atmosphere by a fan, the indoor radon could be maintained below the standard (4pCi/l). After that, the Watras family moved back into that house and lived there. However, based on the risk assessment of the Environmental Protection Agency, doctors warned the family that they were 7 times more likely to die from lung cancer within 10 years than the general public who were not exposed to radon. Moreover, they said that the three children (then a 1-year-old girl, a 3.5-year-old boy, and a 6-year-old boy) could die before they even become adults. The Watras family had to live in fear of eventually dying from lung cancer because they lived in a high radon concentration house for about one year.

As a result of this incident, the Environmental Protection Agency investigated the radon status of the entire United States and its houses. Radon was emitted from the geology of all states, and about 6% of the houses surveyed exceeded the safety standard. It was also estimated that about 21,000 people, equivalent to 13% of the annual lung cancer deaths (about 160,000 people) in the United States, were due to radon, and radon was found to be the second highest cause of lung cancer after smoking. Therefore, the American Medical Association and the Environmental Protection Agency suggested radon testing for all houses. The Environmental Protection Agency regulated by law that radon should be included in the information disclosed to buyers during house transactions across the United States, and the National Association of Realtors (NAR) also joined in 1993.

Thirty years after the incident, in 2015, another interesting fact was revealed to the media. The Watras family had been living in that house without migrating elsewhere. However, contrary to what the doctors had warned at the time, no one they did not get lung cancer. The couple had no abnormalities, and all their children grew up healthy and became adults. The prediction 30 years ago was completely off. Rather, it is said that the Watras family lived in severe mental pain due to the wrong prediction. At the time of the incident, the doctors made exaggerated predictions without accurate information about the extent and duration of the Watras family's radon exposure.

It was already over a hundred years ago that radon was revealed to be a harmful radioactive substance to the human body. Also, the fact that it is a major cause of lung cancer was made known by miners working in mines in the 1950s. However, nowhere in the world has it been reported that high concentrations of radon gas can be observed in homes. This incident left a new environmental issue of 'harmfulness of natural radioactive substances and a safe home from radon' in various countries around the world, including the United States. And it became an opportunity to establish the Home and Building Inspection and Certified Building Biologist system in major developed countries in the United States and Europe. Building biologists are professionals whose occupation originated in Germany. Their goal is to create healthy homes, schools, and workplaces by ensuring the supply of clean indoor air and tap water free from pollutants, and by eliminating electromagnetic radiation.[2] The U.S. Environmental Protection Agency has designated January, when home ventilation is relatively likely to be harmful, as 'National Radon Action Month' and is raising awareness among the public.

[2] The Non-Tinfoil Guide to EMFs: How to Fix Our Stupid Use of Technology, Nicolas Pineault, CreateSpace Independent Publishing Platform, 2017.

Chapter 36
Humidifier Disinfectant Catastrophe

In April 2011, pregnant women showing severe lung disease symptoms were consecutively admitted to the emergency room and respiratory department of a university hospital in Seoul. The medical staff were looking for other causes because no deadly bacteria or viruses were found from the patients during the examination. A 34-year-old female patient died on May 10, and three women died consecutively in June. During the same period, seven infants also died from the same symptoms.

As similar cases were reported nationwide, the Korea Disease Control and Prevention Agency conducted an epidemiological investigation. The first results were announced in June 2011, the main content of which was that the patients had inhaled 'something' in the air, causing damage around the bronchi, and all the deceased showed symptoms of pulmonary fibrosis, where the lungs harden. Two months later, they added the conclusion that "the cause of the lung disease is presumed to be a humidifier disinfectant."[1]

However, the government could not ban the sale of humidifier disinfectants. They were collecting products distributed in the market and conducting animal inhalation experiments, but there were no related legal provisions, so it was difficult to ban sales or force recalls until clear results were obtained. The Korea Consumer Agency merely recommended refraining from using humidifier disinfectants on September 30. It was not until November 11 that the Ministry of Health and Welfare issued a collection order for humidifier disinfectants confirmed to be harmful based on the epidemiological investigation results of the Korea Disease Control and Prevention Agency.

In January 2012, victims began to file civil and criminal lawsuits against humidifier disinfectant sellers and the state. In February, the Korea Disease Control and Prevention Agency finally confirmed that the humidifier disinfectant was the cause of lung damage based on the results of animal inhalation experiments. The toxic

[1] https://en.wikipedia.org/wiki/south_korean_humidifier_disinfectant_case

S. S. Park, *Eco-disasters*, https://doi.org/10.1007/978-981-96-8667-4_36

substances included in the humidifier disinfectant were Polyhexamethylene guanidine (PHMG) and Oligo(2-)ethoxyethyl guanidine chloride (PGH), and some products also used Methylchloroisothiazolinone (MCI, or MCIT). These chemicals were biocides mainly used for purifying or cleaning water tanks. They were used in various products such as shampoo and wet wipes, not just humidifier disinfectants, because their skin toxicity was only about one-tenth of other disinfectants.

The problem was that these biocides were used as humidifier disinfectants that go into indoor air. Ultrasonic humidifiers commonly used in homes and offices send fine water droplets into the air. At this time, the chemicals included in the humidifier disinfectant These particles, equivalent to one-hundredth the size of a water droplet, float in the air and penetrate deep into the lungs when inhaled, damaging the alveoli. The damaged alveoli try to recover, but repeated damage and recovery lead to pulmonary fibrosis, where cells harden. The lungs, unable to function properly, can lead to respiratory distress and, in severe cases, death.

The humidifier disinfectant was developed in 1994 and included in 24 products for 17 years until it was banned from sale in 2011. A total of 7.19 million units were sold. It is estimated that between 3.5 and 4 million people have used the disinfectant. The government has been conducting damage surveys in five stages since November 2011. As of February 2023, a total of 7,822 victims have applied, with 1,810 deaths. In 2017, the government enacted and implemented a special law for the relief of victims of humidifier disinfectants and their bereaved families.

The number of victims of this incident is estimated to be much higher than the number revealed so far. According to a study released in 2016, 20% of Koreans were exposed to humidifier disinfectants, and it was estimated that 20,000 of the 70,000 pneumonia deaths investigated in Korea from 1995 to 2011, or 29%, were due to humidifier disinfectants. Many people have suffered from lung diseases or died from them since the first sale of disinfectants in 1994, and the number reported since 2011 is only a fraction of the victims.

The worst chemical disaster, in which about 20,000 people died, was published in 2016 as a novel titled "Bacteria: Humidifier Disinfectants and Untold Things" and made into a movie titled "Air Murder" in 2022. It tells a sad story of so many innocent citizens who lost their lives to crimes committed by companies that took advantage of the government's negligence in chemical management and those who survived have to live in pain for a lifetime.

Modern humans live in a flood of chemicals. Many of these are household chemicals that can always come into contact with and be inhaled by the human body through living spaces. Rapidly developing science constantly produces new substances that are marketed with government approval. However, the general public knows little about these substances. The public simply uses them knowing only that the government has approved their use. There is a blind spot here. The government's management of chemicals is extremely lax. The humidifier disinfectant incident is a stark illustration of the problems with household chemicals in modern society.

There are three main routes through which substances can enter the body from the outside. These are the digestive organs (Ingestion) through water and food, the respiratory organs (Inhalation) through air, and skin contact. The same substance at the same concentration can enter the body through these routes. Even if it enters the body, the effect on the human body can be very different depending on the route. The problem with humidifier disinfectants is that they entered the respiratory system. The government allowed the use of PHMG, PGH, MCI, etc., which were in the disinfectant, without testing the toxicity that appears when inhaled into the respiratory system. In particular, humidifier disinfectants were classified as industrial products, so only the general safety standards under the 'Quality Management and Industrial Product Safety Management Act' were applied, failing to prevent damage.

In response to this incident, the South Korean government enacted the 'Act on Registration, Evaluation, etc. of Chemicals' in 2013. It established a new legal device to protect public health and the environment from chemicals. However, it is still full of loopholes. Currently, the Korean government is applying a prohibition regulation (Negative System) that prohibits the use of certain harmful chemical items. On the other hand, in Europe and the United States, the government applies a usage regulation (Positive System) that only allows the use of certain safe chemical items. In the usage regulation, there is no possibility for the general public to use harmful chemicals because the government manages up to the producer's items. On the other hand, in the prohibition regulation, there is a sufficient possibility for the general public to use harmful chemicals that are not designated by the state because the management is delegated to the producer.

James G. Speth, a representative international environmental policy expert in the United States and a professor at Yale, wrote in his book "Red Sky in the Morning (2004)" that environmental education should be mandatory for all students discharged from schools as one of the eight major transitions in modern society. The reason is to reduce the gap between rapidly developing science and technology and the stagnant knowledge of the general public. There are health hazards everywhere in the living environment created by modern civilization, but he argues that modern people living there cannot understand this with their level of knowledge and can only lead to accidents and suffer damage. In other words, environmental education is needed for survival. Lack of environmental knowledge means that it is impossible to live a healthy and pleasant life in modern civilization.

Chapter 37
San Diego Electromagnetic Radiation Mystery

A series of brain tumor deaths among faculty and students occurred at San Diego State University in California, USA. The brain tumor deaths, which began in 1993, increased even more in the 2000s. Notably, all seven deceased staff members (1 in 1993, 4 in 2008, 1 in 2010, 1 in 2013) and three students (1 in 2008, 1 in 2011, 1 in 2012) had worked in or near Room 131 of Nasatir Hall or adjacent buildings on campus.

This mystery became known to the world when Virginia Faver, the mother of Richard Faver, a graduate student in political science at the university who died in 2008, publicly traced the cause and process of her son's brain tumor. Especially, her records were introduced in a book published in the United States in 2014, "An Electronic Silent Spring," and this became a representative case to inform the harmfulness of Radio Frequency (RF) electromagnetic radiation.

The protagonist of the story, Richard, started his master's program in political science at San Diego State University in the fall of 2005. The healthy young man in his early twenties experienced headaches, short-term memory loss, loss of balance, cerebral hemorrhage, seizures, and eventually was diagnosed with 'Glioblastoma Multiforme (GBM)' in March 2008. The doctors who diagnosed and treated his disease suggested that his brain tumor was not genetic and that the use of a mobile phone was the cause. It was revealed that Richard used his mobile phone for 2–3 h every day from 2001, when he turned 22, until 2008, when he started having seizures.

Immediately after his diagnosis, Richard was transferred to the 'Tisch Brain Tumor Center' at Duke University in North Carolina, a prestigious institution in the field of brain tumors, where he received all treatments and surgeries. However, he lost his sight, became paralyzed, and eventually passed away on October 11, 2008, seven months after his seizures began. U.S. Senator Edward Moore 'Ted' Kennedy also received treatment for the same brain tumor here and died on August 25, 2009.

Richard's mother, who was devastated by her son's death, began investigating the relationship between brain tumors and cell phone use a few months later. She thought that finding and sharing that relationship would make her son's death not in vain, and

S. S. Park, *Eco-disasters*, https://doi.org/10.1007/978-981-96-8667-4_37

she didn't want any other parents to go through the hard pain they had experienced. She also believed that Richard would want the real cause of his death to be conveyed to his colleagues and friends.

As she was gaining confidence about the relationship between cell phones and brain tumors by investigating numerous research results, she came across new news. Around July 2009, she saw an article about a group brain tumor occurrence that had happened at San Diego State University. The article was about four faculty members who spent a lot of time in Nasatir Hall Room 131 all having brain tumors. Room 131 is where the staff of the Political Science Department at San Diego State University used as an office and lounge. Richard also spent seven days a week in Nasatir Hall Room 131 while working as a teaching assistant. The article reported that there was a relay antenna installed in the communication building next to Nasatir Hall, and three students, including Richard, also died.

She began investigating additional victims related to the relay antenna tower next to Nasatir Hall. The tower is the Ground Wave Emergency Network (GWEN) installed at the highest point in the campus. The radio signals from the antenna of this tower are used for police, ambulances, homeland security networks, and were also needed for cell phones and WiMax (Worldwide Interoperability for Microwave Access). In the summer of 2005, around the time Richard was admitted, the antenna of San Diego State University was upgraded with advanced equipment. The signal emitted by the antenna covers a radius of about 116 km and can pass through almost all structures. A dish antenna was also installed on the tower for the High Performance Wireless Research and Education Network (HPWREN).

She was convinced that her son developed a brain tumor because he used his mobile phone for 2–3 h a day and lived in a building next to the relay antenna tower. Therefore, she contacted Stephen Weber, the president of San Diego State University, and the board of directors to request a review of the harmful effects of Nasatir Hall. The university commissioned Thomas Mack, a preventive medicine specialist, to review the harmful effects. However, his report stated, "Based on the fact that I have never received a report that people working in this building have been exposed to any abnormal chemicals or radiation, I presume there is no effect."[1] She then contacted Arnold Schwarzenegger, the Governor of California. The governor sent the issue back to the university, suggesting it was an internal problem.

Her efforts ended without any significant income. This incident, which resulted in many casualties, failed to achieve any results in the United States, where science and human rights are ahead of any other country in the world. The harmful effects of the radio frequency transmitted by mobile phones on the human body have been scientifically proven for a long time. Our brains have a blood–brain barrier (BBB) that prevents toxins in the blood flowing through the arteries from entering. If you use a mobile phone for about two hours, the function of the barrier weakens. If the delicate selective function of the barrier weakens, a small amount of harmful substances in the blood vessels can enter the brain and cause a tumor. Therefore, in

[1] An Electronic Silent Spring: Facing the Dangers and Creating Safe Limits, Katie Singer, Portal Books, 2014.

2011, the International Agency for Research on Cancer (IARC), under the World Health Organization (WHO), classified mobile phone radiation as a group that can cause cancer in humans. This group includes lead, DDT, asbestos, chloroform, etc.

A similar case occurred at the University of California at San Diego campus.[2] At least eight workers in the Literature Building have been diagnosed with breast cancer since 2000 and two died. In addition, a handful of other cancers, including ovarian and salivary gland cancers, have also been reported. Due to the high incidence of breast cancer, requests for an investigation were made in 2002 and 2005, but it was not until 2007 that an investigation began. The investigation concluded that there was no "possible evidence" that could explain the increased cancer rates.

However, students, professors, and staff argued that the university failed to address concerns that the Literature Building had an alarmingly high number of cancer cases, protesting with signs that read "crisis of safety". This pressure led the university to conduct a further investigation in 2008, and both elevators in the building, suspected of generating carcinogenic electromagnetic fields, were shut down. The new investigation concluded that "there may be a slight or modest increase in breast cancer risk associated with a small area on the first floor very close to the electrical room and the elevator room." The university "evacuated areas at risk of exposure to high EMF (electromagnetic fields) levels" and replaced the elevator motors and other equipment to reduce the generation of EMF.

The harmfulness of the radio frequency radiation used by various wireless communication devices, including mobile phones, is now The government's regulation is too inadequate and users are unaware. So even when such incidents occur, the result ends in vain. This is the reality. The scientific debate continues in the United States and Europe about the harmfulness of electromagnetic waves generated from wireless communication technology and electricity use.

[2] https://archive.nytimes.com/well.blogs.nytimes.com/2009/02/24/campus-building-blamed-for-cancer-cluster/

Part VII
Cross-Border Disputes and Planet in Danger

Chapter 38
Scandinavian Acid Rain

The Scandinavian Peninsula, where Northern European countries such as Sweden, Norway, and Finland are located, occupies more than a quarter of the total area of Europe, yet its population is only about 30 million. These countries, with most of their territory consisting of mountains and lakes and a stable social structure prioritizing peace and welfare, are relatively distant from environmental pollution. However, signs of abnormalities began to appear in the forests and lakes here from the 1950s. The once lush forests gradually disappeared, and the fish in the rivers and lakes began to vanish. This was due to acid rain caused by air pollutants emitted by neighboring countries.

The Scandinavian Peninsula has a long history of suffering from air pollution damage from neighboring countries. "Brand", published in 1866 by Henrik Ibsen,[1] a world-renowned Norwegian playwright well known to us as the Doll's House, features a pastor who falls down while searching for an ideal and says, "Britain's creepy coal clouds are floating low, covering the whole country black, dirtying the fresh green and spewing poison." This indicates that people here knew from the nineteenth century that the heavily polluted British atmosphere due to coal usage was causing damage to the Scandinavian Peninsula.

The countries of the Scandinavian Peninsula, which had been suffering damage for a long time, began to publicize the issue of acid rain as one of the solutions. As part of this, the first 'United Nations Conference on the Human Environment' was held in Stockholm, Sweden, in 1972. This conference was an important catalyst for the international community to take interest in environmental issues. At that time, Sweden was the country in Northern Europe most severely affected by acid rain, with about 40,000 of its 90,000 lakes turning into dead lakes where no life could survive. Due to these problems, Sweden established the world's first independent environmental administration, the Environmental Protection Agency, in 1967. This

[1] https://www.amazon.com/Brand-Version-Geoffrey-Penguin-Classics/dp/0140446761

S. S. Park, *Eco-disasters*, https://doi.org/10.1007/978-981-96-8667-4_38

was three years ahead of the establishment of the Environmental Protection Agency in the United States and the Environment Agency in the United Kingdom in 1970.

Acid rain was first reported by British scientist Robert A. Smith in 1872,[2] but the severity of the damage was not known until 1955 when British scientist Eville Gorham investigated the causes of acid rain and its impact on ecosystems.[3] The causes of acid rain include sulfur oxides and nitrogen oxides produced when coal or oil is burned. Generally, rain or snow contains carbon dioxide in the atmosphere, so it shows a slight acidity. Therefore, pure The pH of water is 7.0, but the pH of rainwater falling from the sky is about 5.6. However, when coal and oil are burned, the sulfur oxides and nitrogen oxides that are emitted stay in the atmosphere and become strongly acidic when they enter clouds or seep into falling rain or snow. These oxides also create acidic fog or frost and are adsorbed onto dust and fall to the ground. All forms of precipitation that come down from the atmosphere with acidity are collectively called acid rain.

Damage caused by acid rain manifests in various ways, including deforestation, reduced crop yields, death of organisms living in lakes and rivers, and property damage. Acid rain kills soil microorganisms, inhibits organic matter decay, and causes the loss of trace elements necessary for plant growth, such as potassium, calcium, magnesium, and sodium contained in the soil, reducing fertility. Also, when the water in lakes and rivers becomes acidic, aluminum in the sediment or soil is leached out, killing aquatic life. Acid rain thus makes the soil barren and leaches toxic substances into the water, making it uninhabitable for life, and deforms the earth's surface, which is why it is also called Earth Leprosy, a disease of the earth created by industrial civilization.

Strong acid rain can cause skin diseases or eye diseases in humans. Also, if acidified water is used as a drinking water source, the water pipes can corrode, causing fatal damage to the health of residents. For example, in some regions of Sweden where acid rain damage is severe, there have been reports of cases where high concentrations of copper ions were leached from drinking water pipes, causing the hair color of residents to turn red. Acid rain dissolves metals, cement, marble, granite, etc., causing a lot of property damage by corroding vehicles, buildings, and monuments made of these materials. Europe's Notre Dame Cathedral, Cologne Cathedral, and Greek monuments in Athens have already undergone significant corrosion due to acid rain.

Acid rain damage is closely related to soil composition. If the soil contains a lot of lime and has a thick topsoil layer, it can withstand acid rain, but if the opposite is true, the damage can be very severe. The reason why most of the lakes in the Scandinavian Peninsula have turned into dead lakes and there is significant damage to the forests is because the soil here is particularly sensitive to acid rain. On the other hand, in the case of southern Europe, Greece, where the soil contains a lot of lime, It neutralizes acid rain well, resulting in relatively less damage. According to one study, the soil in Greece was found to be 10 times stronger against acid rain than the soil in the Scandinavian Peninsula.

---

[2] https://en.wikipedia.org/wiki/robert_angus_smith

[3] https://en.wikipedia.org/wiki/eville_gorham

Environmental disasters caused by acid rain are not limited to certain areas, but are widespread across the globe. In particular, the damage from acid rain in Europe and North America is very serious. Not only in the Nordic Scandinavian region, but also in Germany and Switzerland, 50% of the forests have been devastated by acid rain, and hundreds of thousands of hectares of forests have died in the Czech Republic and Poland. The forests in the eastern United States and Canada have already lost their original appearance over a wide area. It was once known that 20% of the lakes in the northeastern region centered around New York State in the United States had been acidified.

According to the analysis of satellite photos observed in the 1990s, it was found that acid rain regularly falls in a vast area from the UK to Central Asia. Also, satellite photos showed that acid rain generated in North America crosses the Atlantic and moves to Europe. Air pollutants emitted in the Northern Hemisphere gather towards the polar regions, while pollutants in the Southern Hemisphere do not travel far and mostly stay in the area where they occur.

National conflicts due to acid rain are appearing not only in the Scandinavian Peninsula but also in various parts of the world. Canada claims that more than 70% of the substances causing acid rain in their country come from the United States, and it has long been internationally recognized that most of the causative substances in the Scandinavian Peninsula come from the UK and Germany. Japan has announced that the contribution to acid rain in their country is 50% from China, 15% from the Korean Peninsula, and 35% from within Japan. It is estimated that 30–50% of the cause of acid rain in Korea is due to air pollutants from China. China, the world's largest producer and consumer of coal, is still known to have acid rain falling on 30% of its territory. Neighboring Korea and Japan are suffering from acid rain damage as sulfur dioxide flies in with fine dust on the westerly winds. Generally, acid rain travels on low currents in the troposphere, so the offending and victim countries are geographically divided. In other words, Portugal, Ireland, the UK, Germany, China, and the United States become offenders, and the Scandinavian Peninsula, Poland, Korea, Japan, and Canada become victims.

If we look at the research results analyzing the historical changes of substances causing acid rain using the glaciers of Greenland, we can see that there have been two maximum accumulation points of sulfur dioxide so far. One was from 1875 to 1910, when a relatively large amount of sulfur dioxide was emitted, and the other was from 1945 to 1975, when the emission was much higher than before. This means that a large amount of sulfur dioxide was emitted into the atmosphere due to rapid industrialization after World War II. On the other hand, nitrogen oxides have shown a rapid increase since 1950 when looking at the records accumulated in glaciers. This is closely related to the increase in automobile use.

The 1972 Stockholm UN Human Environment Conference, held in the wake of acid rain damage in the Scandinavian Peninsula, played a significant role in raising awareness of the seriousness of environmental issues and calling for international cooperation. In the case of North America, the amount of acid rain-causing substances began to decrease after 1975 in industrial areas around the Great Lakes and the Ohio River, which once emitted enormous amounts of acid rain-causing substances, and it is known to have decreased by about 15~25% so far. Europe was not successful in reducing the causative substances initially. Continued industrialization led to an increase in energy use from 1950 to 1980, and the rapid increase in cars led to a steady increase in acid rain-causing substances. However, substantial efforts to reduce acid rain began with the signing of the 'Helsinki Protocol' in 1985 for the reduction of sulfur dioxide emissions and the 'Sofia Protocol' in 1988 for the reduction of nitrogen oxides. In fact, Europe's sulfur dioxide emissions in 1995 were 15% lower than in 1980. This was mainly the result of improved energy efficiency. However, the reduction of nitrogen oxides through the restraint of car use was not very successful. Until 1980, the contribution of sulfur dioxide and nitrogen oxides to acid rain in Europe and the United States was about 2 to 1, but since then, the contribution of sulfur dioxide has continued to decrease and is now close to 1 to 1.

Chapter 39
Jordan River Water War

The Jordan River, originating in Syria and flowing through Jordan and Lebanon to Israel, is the lifeline of the Middle East. On the dawn of June 5, 1967, the water war, which signaled the global water crisis, began here. When Syria blocked the waterway upstream of the Jordan River, Israel launched a full-scale attack, bombing Syria's construction equipment and ending the war in just six days.

This war, also known as the Third Middle East War or the Six-Day War, ended in a complete victory for Israel.[1] Israel gained control over the groundwater in the upper Jordan River and the West Bank area. Syria's allies, Egypt and Jordan, also participated in the war but were defeated. The Israeli Air Force bombed over 400 Arab alliance aircraft within three hours of the outbreak of war. Of these, 286 were Egyptian aircraft, and only 19 Israeli aircraft were destroyed. The Israeli forces completely seized air superiority at the outbreak of the war and advanced to the Suez Canal through the Sinai Peninsula. Egypt, the leader of the Arab alliance, accepted the UN's ceasefire recommendation just four days into the war after most of its military equipment was destroyed by Israel's surprise attack. Syria, which had initially sparked the conflict, stood by idly once the war began, ultimately ending up in a position of surrender.

Later in 1994, a peace treaty was signed between Israel and Jordan, and an agreement was reached to resolve the water dispute. This treaty appropriately addressed various issues related to water in the Jordan River basin, including the distribution of water between countries and the sharing of related data. The treaty also included a comprehensive plan to jointly manage and develop the water resources of the Jordan River. Both governments came to believe that only mutual cooperation could overcome water shortages.

However, in the twenty-first century, clouds of war are gathering again over this area due to water. This time, a water dispute has started between Israel and Lebanon. The conflict began in the spring of 2001 when Lebanon started diverting water from

[1] https://en.wikipedia.org/wiki/Six-Day_War

S. S. Park, *Eco-disasters*, https://doi.org/10.1007/978-981-96-8667-4_39

the Hasbani River, which flows into Israel's Sea of Galilee, to nearby villages. Israel claims that the Hasbani River accounts for 20–25% of the water in the Sea of Galilee. Furthermore, Israel argues that it cannot tolerate Lebanon diverting water because the Sea of Galilee, Israel's largest water source, is increasingly drying up. Lebanon, on the other hand, firmly rejects Israel's claims, questioning how much water could be diverted through a 10 cm diameter pipe. Lebanon asserts that it is merely providing a small amount of water to drought-stricken villages near the Hasbani River and accuses Israel of overreacting. Israel has warned that any neighboring country diverting its water could lead to war at any time.

Another region with a high possibility of water wars is the Nile River basin, which runs through eight countries in North Africa. Egypt, which gets 90% of its water resources from the Nile, is always in a tense relationship with Sudan and Ethiopia upstream. Egypt is ready to wage war at any time if these countries weaponize the water of the Nile. In 1978, then Egyptian President Anwar Sadat said that if any country tries to take away Egypt's lifeline, it will have no choice but to go to war. Boutros Ghali, a former UN Secretary-General who served as Egypt's foreign minister, also predicted that the next war would be sparked by the water issue of the Nile.

Egypt already signed an agreement on water distribution with Sudan upstream of the Nile in 1959. This agreement stipulates that Egypt uses 55.5 billion cubic meters of water per year and Sudan uses 18.5 billion cubic meters. However, this agreement only contains the distribution of water between Egypt and Sudan, and other basin countries such as Ethiopia, Uganda, and Kenya are excluded, which is a major cause of conflict. Especially, the explosive population growth in these countries is leading to a huge demand for water, and conflict is expected to be inevitable soon.

In addition, Turkey, Iraq, and Syria are in conflict over the Tigris and Euphrates rivers. One side's economic development plan worsens the other side's water supply situation. Hungary and Slovakia in Europe are also arguing over the change of the Danube River's waterway, and India and Bangladesh in South Asia are in conflict over securing agricultural water from the Ganges River. Countries with a high risk of water conflict are mostly dependent on other countries for the water they need. For example, Egypt uses 97%, Cambodia 82%, Syria 79%, Sudan 77%, and Iraq 66% of water coming from other countries. However, countries where international cooperation on river management and water use is well established are not like this. A representative example is the seven countries in the Rhine basin, where the Netherlands at the end depends on 98% of the upstream countries, but they live without conflict or inconvenience. It produces even the world-renowned beer Heineken with its outstanding water treatment technology.

Countries that have experienced water disputes say that water has a much stronger potential to cause wars than oil because water secures the lives of their citizens. If there is no oil, other alternatives can be used, but there is nothing that can replace water to sustain life. Humanity has lived for hundreds of thousands of years without oil, and there are many places that still live independently of oil. However, no country can withstand without water, so the risk of war due to water is that much greater.

On Earth, a tremendous amount of water covering 70% of the Earth's surface reaches 1.4 billion km^3. However, most of the water is saline that cannot be used for drinking, and only 2.6% is freshwater without salinity. Two-thirds of this freshwater is trapped in glaciers in the Arctic and Antarctic, and one-fifth of the remaining water is in areas that cannot be used or falls at the wrong time and place, leading to disasters like floods. In the end, the freshwater resources that humans can use are only about 1 million tons, or 0.0072% of all the water on Earth.

If all of humanity on Earth evenly uses this freshwater resource, it can be sufficiently supplied to everyone. Unfortunately, however, freshwater resources are not evenly distributed across the globe. Some countries have a tremendous amount of water, while others do not. If you investigate the freshwater resources available for annual use per capita, there is a big difference between countries. Iceland is the world's richest water country with 606,500 tons per capita, and Canada (94,000 tons), Norway (88,000 tons), Russia (29,000 tons), Sweden (20,000 tons), and the United States (8,900 tons) also have relatively abundant freshwater resources. However, Kuwait is the world's worst water-poor country with 11 tons, and Israel (289 tons), the United Arab Emirates (64 tons), and Egypt (43 tons) are also experiencing serious water shortages.

During the last century, the explosive population growth and industrialization on Earth have greatly increased water usage. In just a century, the world's population has quadrupled, the economy has grown 14 times, industrial production has increased 40 times, and cultivated land has increased fivefold, resulting in a ninefold increase in water usage. This means that water usage has doubled every 20 years in the last century. Moreover, the increased water usage has led to more wastewater polluting rivers and lakes. Given the current population growth trend, it is estimated that global water usage will increase by approximately 40% by 2020, and the water needed for food production alone will increase by about 17%.

Due to water shortages, pollution, and lack of infrastructure, 40% of the world's population in over 80 countries are suffering from water-related issues, and the UN reports that over 5,000 children die every day due to water-related causes. Particularly in Africa, India, and China, millions of people die each year due to water-related issues. In response to the recommendations of the Rio Environmental Summit held in June 1992, the UN declared March 22 as 'World Water Day' at its 47th General Assembly in November of the same year.

In the last century, the world waged several wars to obtain oil. However, in the twenty-first century, major international organizations, including the World Bank, foresee that wars to obtain water will be more frequent than those for oil. If the twentieth century was the century of fire (oil), the twenty-first century is the century of water, and water will soon determine the future of the Earth and humanity. Currently, about 40% of the world's population lives on water coming from neighboring countries, and water disputes between countries are already occurring in over 30 countries. The 214 multinational rivers flowing on Earth could become powder kegs that could ignite a war at any time if the water situation worsens.

Chapter 40
Koko Toxic Waste Dumping

An Italian working at a construction company in Nigeria conspired with a Nigerian to smuggle a total of 38,840 tons of hazardous waste from Italy disguised as chemical products into the Koko Port in Bendel State, Nigeria, over five occasions from August 1987 to May of the following year.[1] The waste barrels contained hazardous chemical materials such as a dioxin-producing combustible known as polychlorinated biphenyl (PCB), as well as dimethyl formaldehyde and asbestos fibers. In Koko, the corroding barrels were stored in the backyard of an unsuspecting farmer. After a few months, they started leaking and contaminated soils and groundwater, making several villagers seriously ill.

When it was revealed that the cause of the damage was industrial waste from another country, the Nigerian government denounced Italy's unconscionable act to the international community. In response to the strong reaction of the Nigerian government, the hazardous waste was expelled and shipped back overseas. The ship carrying the waste attempted to dock in Spain, France, Belgium, the Netherlands, Germany, and the United Kingdom from July to August 1988, but no country accepted it. Eventually, the hazardous waste returned to Italy through international public opinion and several rounds of diplomatic negotiations by the Nigerian government. However, the Nigerian government had to bear the costs of cleaning the contaminated area and treating the affected residents.

This act of developed countries sending their hazardous waste to developing countries is called Third World Dumping, and many African countries were using it as a means of earning foreign currency at the time. According to statistics, the amount of hazardous waste that crossed borders from Europe in one year in 1985 was a whopping 3 million tons. Not only Europe, but also the United States and Japan sent large amounts of industrial waste overseas. However, in the wake of the Koko Dumping Incident, Nigeria appealed to several African countries not to import industrial waste from other countries in order to protect their territory and people from environmental

[1] https://www.environmentandsociety.org/perspectives/2013/1/article/contribution-cartoonists-environmental-debates-nigeria-koko-toxic-waste

S. S. Park, *Eco-disasters*, https://doi.org/10.1007/978-981-96-8667-4_40

pollution. They also urged developed countries to stop such exports that only seek their own interests. In the same year, the Organization of African Unity urged its member countries to ban the import of industrial waste from other countries. The World Bank also declared that it would not provide loans for projects related to the international trade of hazardous substances.

In the past, developed countries had strict legal regulations on the disposal of hazardous waste, whereas developing countries had very weak regulations. This made it easy to import hazardous waste from developed countries into developing countries. In developing countries that accepted hazardous waste, there was a lack of not only legal regulations but also knowledge and management techniques about the dangers of this waste. Therefore, the dangers of hazardous waste are greatly amplified in the developing countries that accept it compared to the countries that emit it. In fact, Nigerians living in the Koko area, who did not even know what hazardous waste was, suffered serious health damage when local people were mobilized to collect the abandoned waste. About 150 Nigerians who were collecting the waste showed symptoms of nausea, hemoptysis, and paralysis due to toxic chemicals leaked from the waste, and some were hospitalized in a coma with burns.

A similar case is the Khian Sea incident in the United States. In August 1986, the Khian Sea ship traveled to various ports carrying ash from a waste incinerator in Philadelphia, USA, and arrived in the Republic of Haiti in the West Indies.[2] This ash was hazardous waste containing cadmium and mercury. The Khian Sea was caught while deceiving it as fertilizer and loading it on the dock, and the Haitian authorities ordered the cargo to be reloaded. However, the Khian Sea is known to have dumped 2,000 to 4,500 tons of ash near Haiti while reloading it. After that, the ship wandered around West Africa for over a year, crossed the Indian Ocean through the Suez Canal, and finally arrived in Singapore. But at this time, the ship's cargo hold was empty. The outrageous act of the Khian Sea could not avoid international criticism.

As the movement of hazardous waste between countries frequently occurred and became a global issue, the need for an international convention to regulate this began to be raised. Then, after the Koko dumping The incident directly led to the adoption of an international convention to regulate the transboundary movement of hazardous wastes in Basel, Switzerland on March 22, 1989. This convention, known as the 'Basel Convention', came into effect in 1992.[3] The main content of this convention is to specify the types of waste subject to regulation and to establish conventions and responsibilities related to import and export, and to obligate prior notification and consent during the movement of hazardous waste from importing countries, exporting countries, and transit countries. The ultimate goal of the Basel Convention is to minimize the generation of hazardous waste and to handle the hazardous waste generated in the country itself. In particular, this convention strictly regulates the sending of hazardous waste to countries that do not have the ability to dispose of it in an environmentally friendly way.

---

[2] https://en.wikipedia.org/wiki/khian_sea_waste_disposal_incident

[3] https://en.wikipedia.org/wiki/basel_convention

The attempt to bring Taiwan's nuclear waste into North Korea, which became a big issue in the international community in 1997, is also subject to the Basel Convention. At that time, Taiwan and North Korea planned to bury a total of 200,000 barrels of nuclear waste in a closed mine in Pyongsan, Hwanghae Province over two years.[4] This mine was exposed to groundwater pollution because the tunnels were developed unplanned, and it was a place with a high risk of earthquakes due to its geological structure. Also, if 200,000 barrels of nuclear waste were transported by ship, there could have been unexpected accidents at sea. However, both North Korea and Taiwan did not join the Basel Convention, so there was no means to regulate it internationally. Fortunately, this plan did not come to fruition due to strong opposition from the international community and North Korea's inability to complete the nuclear waste storage facility that Taiwan demanded due to economic difficulties.

Similar to third world dumping, there are cases where harmful chemicals produced in developed countries are exported to underdeveloped countries. In developing countries, the lack of knowledge about the substance and the lack of proper regulation can naturally increase the potential for damage to the human body and the natural ecosystem due to misuse and abuse by users. Also, some pesticides are non-degradable and highly toxic, so they are produced in developed countries and exported to developing countries, even though they are banned for use in their own countries. Ironically, developed countries are importing agricultural products produced using these pesticides in developing countries. In the end, developed countries end up bringing back the domestically banned pesticides they produced and sold overseas through imported agricultural products.

To solve these problems, a convention on the international trade of hazardous chemicals and pesticides was adopted in Rotterdam, the Netherlands in 1998.[5] The purpose of this 'Rotterdam Convention' is to prevent environmental and human body damage from using hazardous substances without sufficient knowledge about their harmfulness in the importing country. The aim is to minimize the damage that will be caused. According to this agreement, countries exporting hazardous chemicals must fully convey relevant information to the importing country and obtain prior consent, adhering strictly to the principle of Prior Informed Consent (PIC). The Rotterdam Convention was adopted in 1998 and came into effect in 2004. This convention regulates a total of 37 pesticides and industrial hazardous substances.

Another international convention to regulate the movement of hazardous substances between countries was adopted in Stockholm, Sweden in 2001. The 'Stockholm Convention' aims to reduce the harm to human health and the environment caused by Persistent Organic Pollutants (POPs), regulating 12 substances, the so-called 'Dirty Dozen', such as dioxin, DDT, and furan. These substances, which take decades to decompose, once released into the environment, can have a global impact, necessitating an international agreement, which came into effect in 2004.[6]

[4] https://www.nytimes.com/1997/02/07/world/north-korea-agrees-to-take-taiwan-atom-waste-for-cash.html

[5] https://en.wikipedia.org/wiki/rotterdam_convention

[6] https://en.wikipedia.org/wiki/stockholm_convention_on_persistent_organic_pollutants

In fact, the movement of hazardous waste between countries has significantly decreased since the Basel Convention. The Rotterdam Convention has imposed stronger regulations on the act of producing hazardous chemicals banned in one's own country and selling them to other countries. However, as seen in the nuclear waste trade between North Korea and Taiwan, international conventions are ineffective if the relevant country does not participate. Even if a country participates in the convention, international environmental crimes that cleverly evade surveillance continue to occur.

Chapter 41
Sahel Desertification

The Sahel region of Africa is a vast grassland area located south of the Sahara Desert, spanning between latitudes 14° to 20° North. It stretches approximately 5,000 km from east to west and is about 300 km wide. This region experiences dry conditions for most of the year, with a short rainy season from June to August, during which around 200 mm of rainfall occurs. The northern part of the Sahel is primarily inhabited by nomadic people, while the southern part is home to agricultural communities.

From the 1960s, as the population increased here, desertification gradually began, and with the addition of drought, this region turned into a barren land where life could not survive. Especially in 1972 and 1973, when the damage reached its peak, hundreds of thousands of people and livestock died. Still, people considered it just a simple drought.[1] However, from 1982 to 1985, a greater disaster struck a wider area, and millions of people lost their lives. The Sahel region, which had no major disasters for thousands of years, experienced two enormous disasters in the late twentieth century.

The desertification of the Sahel region was mainly due to the increase in population, the increase in demand for arable land, the spread of slash-and-burn agriculture, indiscriminate logging for firewood, and excessive grazing of livestock, which devastated the grasslands. In addition, a long period of drought continued, and the area became irrecoverable. When vegetation disappears from the surface, the reflectivity of solar energy increases, the surface cools, and the temperature decreases. As a result, cold surface conditions form dry descending air currents, reducing rainfall and decreasing soil moisture, which accelerates desertification at a rapid pace. Also, soil without vegetation is easily eroded by wind and rain, and the land can no longer supply nutrients or moisture. As desertification progresses, the amount of oxygen in the atmosphere decreases, wildlife disappears, and due to water shortage, crop cultivation becomes impossible, leading to severe food shortages.

On Earth, areas where the soil is dry due to more moisture evaporating than rainfall from the sky occupy one-third of the land area. Some of these are originally deserts

[1] https://en.wikipedia.org/wiki/droughts_in_the_sahel

S. S. Park, *Eco-disasters*, https://doi.org/10.1007/978-981-96-8667-4_41

where vegetation cannot live, and the rest are grasslands where grass grows with little rainfall and livestock and agriculture are practiced. Desertification is proceeding centered on these grasslands. Areas where desertification is progressing suffer from hunger and disease, and yellow dust occurs, causing great damage to neighboring countries.

Currently, desertification is underway in 30 million km^2, which is 19% of the Earth's land area, and about 150 million people are threatened with survival due to desertification. Over 100 countries are experiencing or suffering from the effects of desertification. In Asia, about 40% of arable land is at risk, and in Africa, about 80% is threatened by desertification. The situation is particularly severe in regions such as Africa's Sahel, the Andes in South America, and Mongolia and northwest China in Asia. Even now, the procession of refugees leaving to find new homes continues in these areas.

The causes of desertification can be divided into climatic factors such as long-term drought and human factors such as over-cultivation, overgrazing, and deforestation. So far, climatic factors account for only 13% of the causes of desertification, and the remaining 87% are due to human factors. The cause of desertification is the rapid expansion of farmland to feed the rapidly growing population and supply the resources needed by industrial society, and the indiscriminate deforestation. Such over-cultivation and deforestation devastate the soil and change the climate, eventually turning it into a barren land where life cannot survive.

Once desertification begins, it progresses rapidly due to the mutual amplification effects of vegetation destruction, soil erosion, and local weather change. Therefore, to prevent desertification, it must be done in the early stages when vegetation recovery is possible. In particular, it should start before the soil that supplies the nutrients and moisture necessary for plant growth is eroded, and above all, it is best to prevent it in advance in grassland areas where the possibility of desertification is high.

Desertification due to excessive cultivation has already occurred in the south western plains of the United States in the 1930s. As the demand for food surged with the outbreak of World War I in Europe, American pioneers cut down the grass and trees of the western grasslands and created vast farmland. However, a drought that began in 1931 turned the area into a barren land. This event, widely known as the Dust Bowl, occurred in the vast plains spanning southwestern Oklahoma, Kansas, Colorado, New Mexico, and Texas. Many people had to die of starvation and disease or leave, and the yellow dust from here covered as far as New York in the east. Records show that the steps of the Capitol in Washington DC, the US capital, were also filled with yellow dust. This event, along with the economic depression at the time, speaks of the gloomy 1930s in the United States. The famous novel "The Grapes of Wrath" by John Steinbeck is set against this historical backdrop.

The U.S. government declared this area a national disaster area in 1935 and sent relief funds to aid in recovery. It also established the Soil Conservation Service (SCS) under the Department of Agriculture to implement soil erosion prevention policies. Then, in 1937, President Franklin Roosevelt initiated the Shelterbelt Project to revive the barren plains. This project designated a 160 km wide area from Canada to northern Texas to plant native trees that once lived there. The project, which invested 75 million dollars, was carried out over 12 years. Farmers also planted trees on their farmland fences with government support, and extensive soil work was carried out to restore eroded soil. In addition, fallow management, crop rotation, and new farming methods to prevent soil erosion, such as covering the cultivated surface, began to be disseminated throughout the United States. The desertification that lasted for eight years finally began to recover in 1939. Through this, the United States learned an important lesson that reckless over-farming turns fertile land into barren land.

The main desertification areas in Northeast Asia are China and Mongolia. Especially since the 1990s, the desert in the border area between China and Mongolia has been expanding to the Inner Mongolian Plateau. Currently, about 16% of China's total land area is known to have been desertified due to deforestation, topsoil loss, and sand movement. The situation in Mongolia is even more serious, with 90% of the country at risk of desertification, and since the 1970s, 69,000 square kilometers of grassland have disappeared and the number of plant species inhabiting has decreased by a quarter.

Desertification in China and Mongolia is causing yellow dust damage not only in Northeast Asia but also far away in the North American continent. Yellow dust causes various diseases such as visibility impairment, respiratory diseases, eye diseases, and allergies. The damage is particularly severe for children and the elderly, and pollutants from China's industrialization contained in yellow dust pose a significant threat to public health. In addition, yellow dust also causes damage to aviation and precision industries. In places where computer monitors or TV cathode ray tubes are made, even a small amount of dust can be fatal, so the rate of defective products suddenly increases during the yellow dust season. To reduce yellow dust, afforestation projects have been continued in China and Mongolia, but there is no visible effect. More effective measures are to promote artificial rainfall, irrigation watercourses, and afforestation projects in a comprehensive manner.

The world has come to recognize that the threat of desertification is progressing in various parts of the globe, triggered by the disaster in the Sahel. The UN held the United Nations Conference on Desertification (UNCOD) in Nairobi, Kenya in 1977, established the Plan of Action to Combat Desertification (PACD), and has consistently carried out international cooperation by designating June 17 as 'World Day to Combat Desertification and Drought'. It was a major agenda at the Rio Environment Summit held in 1992, and the 'Convention to Combat Desertification' was adopted in Paris, France in 1994.[2] Inspired by the United Nations Convention, the "Great Green Wall Project" and the "Green Earth Project" to turn barren deserts into arable land are currently underway around the world.[3] The desertification process is caused by the complex interaction of various phenomena such as climate, atmosphere, soil, water, vegetation, and topography, so the success of these projects requires a multidimensional approach involving various fields.

[2] https://en.wikipedia.org/wiki/united_nations_convention_to_combat_desertification

[3] https://en.wikipedia.org/wiki/Great_Green_Wall_(Africa)

Chapter 42
Ozone Layer Depletion

In the last century, humanity has created many technological advancements. These have made human life more convenient and affluent, and greatly expanded our living areas. Among them, refrigerators and air conditioners have completely transformed human food habits and living spaces. This has even changed the distribution of the Earth's population. As the use of refrigerators and air conditioners became widespread, many people who used to live in temperate regions moved to tropical and subtropical regions.

When refrigerators were first developed, they used ammonia gas to achieve a cooling effect, but around the time of World War II, the use of Freon gas, which has excellent cooling ability, began to become widespread. Freon gas was invented in 1928 by an American chemist named Thomas Midgley. It is less toxic and does not combust, so it has a low risk of explosion. This substance dissolves oil but does not react with metals or plastics, and its boiling point is at room temperature, so it easily vaporizes or liquefies. Thanks to these properties, Freon gas was widely used not only as a refrigerant for refrigerators and air conditioners but also as a degreaser and cleaner for machinery, metals, and electronic products, as well as a foaming agent, and a propellant in cosmetics and insecticides. The term "Freon" is the product name of the first company to produce this gas, and its official chemical name is Chlorofluorocarbon (CFC).

This substance was recognized for its unique properties and usefulness from the time it was first developed, but its usage was not high until the 1940s due to the economic depression in the United States and the effects of the World War. By 1950, it was only being produced at a rate of about 20,000 tons per year. However, its usage then exploded, and by around 1970, it had increased to about 750,000 tons per year. Until then, no one thought that such a useful substance could cause tremendous damage to the Earth. It was not until 1974 that research results were announced that Freon gas could destroy the ozone layer in the stratosphere that surrounds the

S. S. Park, *Eco-disasters*, https://doi.org/10.1007/978-981-96-8667-4_42

Earth.[1] Professors Sherwood Rowland and Dr. Mario Molina of the University of California, Irvine, reported to the academic community that because Freon gas is inert, most of it is not destroyed when released and reaches the stratosphere, causing a chain reaction where one molecule of Freon gas reacts with thousands of ozone molecules, destroying the ozone layer in the stratosphere.

Scientists first became interested in the destruction of the ozone layer in the early 1970s due to supersonic The development of airplanes was the beginning. Some scientists claimed that the nitrogen gas and water vapor emitted by airplanes would destroy the ozone layer in the stratosphere, which is used as an airplane route. However, it was later revealed that this does not cause a serious problem worth worrying about.

The ozone layer blocks harmful ultraviolet rays from the sun, protecting life on Earth. Ultraviolet rays are divided into three major areas depending on the degree of penetration through the ozone layer and the impact on living organisms. First, ultraviolet rays (UV-A) corresponding to the wavelength 320~400nm pass through the ozone layer in full and reach the ground, but do not harm living organisms. Some of Ultraviolet rays (UV-B) corresponding to the wavelength 280~320nm reach the ground depending on the state of the ozone layer, and while they have the advantage of producing vitamin D in humans and preventing rickets, if there is too much, it can cause skin cancer. Finally, ultraviolet rays (UV-C) with a wavelength of 280nm or less are completely blocked by the ozone layer and do not reach the ground. This wavelength corresponds to the ultraviolet ray (260nm or less) area that destroys cell genes, so if it reaches the Earth, life cannot survive on land.

Ozone layer destruction by Freon gas reduces the blockage of ultraviolet rays (UV-B) corresponding to the wavelength 280~320nm, allowing more to reach the ground. In this case, skin cancer, cataracts, and keratitis increase in the human body, and the immune system weakens, leading to an increase in infectious diseases. Ozone layer destruction causes more damage to whites, who have weaker sun protection on their skin. If the amount of ozone decreases by 10%, the incidence of skin cancer in whites increases by about 26%. There are also research results that the number of cataract patients worldwide will increase by about 1.6 to 1.75 million per year. It also reduces food production by inhibiting photosynthesis in terrestrial plants, and in aquatic ecosystems, it reduces the productivity of plant plankton, reducing catches. Strong ultraviolet rays accelerate the corrosion and aging of building materials, and intensify air pollution in urban areas by activating atmospheric chemical reactions.

The ozone in the stratosphere performs a beneficial function of blocking harmful ultraviolet rays, but when ozone exists near the ground, it acts as a deadly pollutant to living organisms. Therefore, we call stratospheric ozone 'good ozone' and tropospheric ozone near the ground 'bad ozone'. Tropospheric ozone is the result of a photochemical reaction in which nitrogen oxides, emitted during the combustion of fossil fuels, are broken down into oxygen atoms. The biggest source is car exhaust. It is a cause of photochemical smog and has a strong oxidizing power, causing various respiratory diseases including cancer. High concentrations of ozone

[1] https://en.wikipedia.org/wiki/ozone_depletion

are fatal to people with respiratory diseases, and even healthy people complain of visual impairment, chest pain, headaches, and significantly reduced lung function.

Scientists' concerns began to materialize in the 1980s. In 1982, a hole in the ozone layer was discovered in the Antarctic stratosphere, and the thinning of the Arctic ozone layer was also observed from satellites. Damage to the earth's biosphere due to the destruction of the ozone layer also began to be reported. In particular, it was found that the amount of UV-B reaching the ground had greatly increased.

Although most of the sources of Freon gas are located in the Northern Hemisphere, the reason why the destruction of the ozone layer is severe in the Antarctic is known to be due to the unique weather phenomena of the Antarctic. The Antarctic stratosphere does not properly transport heat from low latitudes (equator) regions, causing the temperature of the stratosphere to drop below –80 °C, and at this time, the stratospheric gas forms solid aerosols. However, when Freon gas attaches to this aerosol solid, the destruction of the ozone layer by active chlorine occurs rapidly. On the other hand, the Arctic stratosphere is influenced by the terrain of the Northern Hemisphere, so heat transfer from low latitude regions occurs smoothly. Therefore, the temperature of the stratosphere in the Arctic does not drop below -80 °C, and aerosols are not formed.

Finally, in 1985, the United Nations Environment Programme (UNEP) led a meeting in Vienna, Austria, where representatives from various countries gathered to discuss measures against the destruction of the ozone layer. Then, in 1987, many countries in the world promised to reduce the use of Freon gas and stop all production and use of Freon gas by 2000 at a meeting in Montreal, Canada.[2] The Montreal Protocol was an international agreement to reduce the production and consumption of Freon gas and halogenated substances, but it was revised in June 1990 in London, UK, to completely stop the production and use of these substances by 2000. Since then, the agreement has been revised and measures have been developed in Copenhagen, Denmark in 1992 and Vienna in 1995. Also, Professors Roland and Molina, who announced the ozone layer destruction mechanism in 1974, received the Nobel Prize in Chemistry in 1995 in recognition of their contributions.[3]

As the United Nations-led meeting on measures against the destruction of the ozone layer proceeded, countries around the world have Development of alternative refrigerants began, and the use of Freon gas significantly decreased after 1988. In particular, the United States, Europe, and Japan, which used to use a large amount of it in the past, reduced it by 75~100% from 1986 to 1994. The reason why they could significantly reduce the use of Freon gas is because alternative refrigerants, Hydrochlorofluorocarbons (HCFC) and Hydrofluorocarbons (HFC), were introduced. Especially, HCFC, which has a weak ozone layer destruction function, was promised to be banned in developed countries by 2020 and in developing countries by 2030 under international agreements. Meanwhile, although the use of Freon gas was banned worldwide in 2009, it is very stable and remains in the atmosphere for a long time because it was emitted in the past, so it is estimated to affect the

[2] https://en.wikipedia.org/wiki/montreal_protocol

[3] https://www.nobelprize.org/prizes/chemistry/1995/summary/

destruction of the ozone layer until 2087. Therefore, the period from 1970 to 2070, when the earth experienced strong ultraviolet rays due to the destruction of the ozone layer by humans, is also called the 'Ultraviolet Century'.

Many new materials created by science in the last century were initially used usefully, but often banned due to environmental issues. Freon gas is one of those substances. What's interesting is the life of Thomas Midgley, who invented Freon gas, and the fate of his other invention, 'leaded gasoline'.[4] Both substances were spotlighted as revolutionary new materials for decades, but later met the same end, and Midgley himself ended his life very ironically. Midgley invented leaded gasoline in 1921 to prevent knocking in car and airplane engines, and it was widely used worldwide. However, due to the harmfulness of lead emitted into the atmosphere, the use of leaded gasoline was banned in countries around the world after 1980 and replaced with unleaded gasoline. And Midgley himself met a strange death at the age of 55 in 1944. He contracted a rare disease called polio in 1941, which paralyzes muscles and the central nervous system. He was so incapacitated that he couldn't even get up from bed on his own. Then he invented a device using a pulley and a strong rope to help him lie down and get up from bed on his own, alleviating his discomfort. Then one day, Midgley died after being accidentally strangled by a rope while using his own invention.

Midgley was a genius inventor. He had as many as 117 patents, served as the president of the American Chemical Society, and received numerous awards. Especially with Freon gas and Lead-free gasoline was a remarkable invention that changed the civilization of the twentieth century. And at the annual conference of the American Chemical Society in 1930, he sought to prove that Freon gas was a groundbreaking substance that was non-toxic and non-flammable, in front of a large audience. He famously inhaled the Freon gas and exhaled it into a flame, an anecdote that remains in the history of 20th-century science. Unfortunately, both Midgley himself and his two inventions met a tragic end. The death of Midgley and the end of his two inventions make us reflect once again on the recklessness of the scientific civilization that humanity achieved in the last century.

[4] https://en.wikipedia.org/wiki/thomas_midgley_jr

Closing Remarks

So far, I have written a book focusing on the causes of environmental disasters, the related developing processes and co-factors, the harms and damages, follow-up measures, similar events, and relevant environmental knowledge while organizing them by event. However, while writing this English edition, I had the opportunity to comprehensively review all the disasters. I compared the causes of environmental disasters, examined how they have changed since industrialization, and investigated pre-industrial environmental disasters adopting the age-old method of comparing and constrasting. Finally, I would like to finish this book by pointing out the misconceptions of environmental disasters along with these summaries and reviews and presenting future prospects and tasks to be recommended and executed.

Causes and Trends

The causes of environmental disasters that occurred in the early stages of industrialization were mostly rooted from the lack of environmental science and technology. Disasters caused by numerous people who flocked to the urban cities without sewage treatment or water treatment facilities are notable events such as the London cholera and the Chicago typhoid. Also, air pollution disasters such as those manifested in Donora, London, and the Los Angeles smog are incidents caused by industrialization and urbanization without appropriate environmental technology or control measures. Through experiencing these disasters, mankind has made innovative environmental science and technology advances such as sewage treatment technology, chlorination of tap water, and exhaust gas filtering and purification systems. In addition, future disasters were prevented by strengthening and enforcing environmental laws and systems.

Another cause of environmental disasters was due to unawareness, ignorance, or illiteracy. Disasters such as those occurred at Minamata, Love Canal, and Goiânia fall into this category. Incidents such as the Yokkaichi asthma, Hinkley chromium, Woburn groundwater pollution, and James River Kepone were also exacerbated by the immorality of those very many perpetrators added to the ignorance of the industries. Disasters such as the Michigan PBB, Three Mile Island nuclear power plant, and Tokaimura radioactive material leakage were caused by simple mistakes by

S. S. Park, *Eco-disasters*, https://doi.org/10.1007/978-981-96-8667-4

unskilled or poorly-trained workers, and there were also intentional disasters caused by means of war, like the Gulf War eco-terror. Marine oil spills like the Amoco Cadiz incident are common environmental disasters that occur thousands of times a year around the world, and it is urgent to reduce both the scale of damage and frequency of occurrence.

The most frequently appearing causative substance or agent in environmental disasters is toxic chemicals. In particular, dioxin is a highly toxic non-degradable substance that arises as a by-product of the chemicals produced, and it has been the cause of several disasters such as Agent Orange in Vietnam, Seveso in Italy, and Times Beach in USA. The worst toxic chemical spill occurred in Bhopal, India, resulting in the immediate deaths of over 20,000 people and nearly 600,000 affected victims. Similar to oil spills, the risk of such accidents lurks around the world, necessitating focused management by countries worldwide in the future.

Legionella, radon, humidifier disinfectant, and artificial electromagnetic radiation cases in Part 6 are incidents that should be noted and emphasized for the healthy living environment in modern times. In particular, the misuse of chemical substances in our daily environment poses a high risk of leading to major disasters like the humidifier disinfectant incident. The international or global environmental disasters discussed in Chap. 7, such as acid rain, water wars, illegal waste dumping, desertification, and ozone layer depletion, should be closely monitored and addressed by international organizations in the future.

Pre-industrial Eco-disasters

Large and small disasters have always been ever-present in human history, and civilization has developed and prospered by overcoming them. Primitive humans began to use clothes, fires, and shelter to survive the cold, and were able to avoid wild disasters by starting their relatively settled life about 10,000 years ago or perhaps even longer in the past. However, settled life and lifestyle have brought a new form of environmental disaster to mankind. Infectious diseases caused by drinking water were repeated, and polluted air made people suffer from respiratory diseases at the time.

Efforts to overcome such environmental disasters can be found documented in historical books. Hippocrates of Greece (BC460~BC370) wrote a book called "On Airs, Waters and Places" to prevent environmental disasters. In this book, he described that spring water or rainwater is better than stagnant water, and also said, "When you go to an unknown city, the most careful thing to look at is the water used by the locals." This book maybe reasonably recognized as the first environmental book in human history.

During the Roman era, drinking water and public health were considered important in order to prevent environmental and health disasters. Around the first century, Rome built nine aqueducts, supplying clean water from unpolluted highlands. The aqueducts were equipped not only with sand and gravel for filtration but also sedimentation tanks to capture and filtered out suspended matters and particulates at regular intervals. As a result, environmental and public health disasters such as waterborne diseases were rare. Vegetius, a military strategist of that time, left a document in 375

AD stating, "Soldiers should not drink stagnant or rotten water, because rotten water is like poison and epidemics break out among those who drink it." The document also reads, "Long-term military stays in one place in summer or fall cause disease by rotting water and dirty drinking water, and it is difficult to solve this without moving the place."

However, with the fall of the Roman Empire, public health became problematic and seemingly irrecoverable. Sewage facilities disappeared, streets were littered with garbage and excrement, and houses were dilapidated. Dirt became a part of daily life and diseases were prevalent. At that time, Christianity viewed illness as a punishment for sin and a means for spiritual salvation in the coming world, so people at that time did not worry much about poor health. Therefore, they did not bathe or wash their clothes. Monks and priests only took baths once or twice a year. Preventing and treating diseases was neither encouraged nor welcomed.

Due to these historical circumstances, the Middle Ages experienced numerous environmental disasters. During the Crusades (1096~1099), the majority of soldiers died of infectious diseases on their way to the Holy Land. Also, the outbreak of the Black Death (1346~1353) resulted in the death of 30 ~50% (about 75~200 million people) of the European population at the time. The root cause was deforestation. As the Little Ice Age began on Earth, the need for firewood for heating led to massive deforestation, which eliminated wildlife hawks, the predators of rodents. As a result, the number of rats, the vector of the Black Death, exploded and occupied unsanitary cities. The Black Death in the fourteenth century was the worst environmental disaster in human history.

In the Renaissance era, modern science began to develop, but the living environment of the general public remained unchanged. Dirty city streets, water pollution, and epidemics continued to persist. The invention of the microscope by Leeuwenhoek in 1665 was a groundbreaking scientific advancement with the potential to combat epidemics, yet it made no immediate contribution. It wasn't until nearly two centuries later, in 1854, during the London cholera outbreak, that the cause was identified and measures could be taken.

Misconceptions About Eco-disasters

Most people fall into two misconceptions about environmental disasters. The first is that "industrialization provided the cause of environmental disasters," and the second is the belief that that "behind environmental disasters lies capitalism's pursuit of a wealthy lifestyle." Therefore, the majority of those leading the environmental movements today continue to blame and oppose industrial capitalism whenever environmental disasters occur.

In the early stages of industrialization, disasters such as the London cholera and Donora smog frequently occurred due to the lack of environmental control measures. Hence, such claim might be reasonable or feasible. However, advanced industrial nations have significantly improved their disaster prevention and response capabilities through the development of science and technology, as well as the strengthening of laws and regulations. Such arguments today are therefore far from the truth. Moreover, as seen in the previous section, environmental disasters occurred frequently before industrialization, including mega-disasters such as the Black Death.

The affluence in capitalist societies has actually reduced environmental disasters. This can be easily understood by analyzing the damaging consequences of the disaster. The degree of disaster damage is determined by the combination of the hazardousness inflicted and the vulnerability of the place being affected. Most of the poor underdeveloped and developing countries are very vulnerable to disasters. Therefore, when the same scale of disaster hazard is inflicted, the damage will inevitably be greatly amplified. This can be confirmed by the fact that more than 95% of deaths from disasters worldwide are currently accounted for by underdeveloped and developing countries.

Future Outlook and Tasks

Disasters have always been with human life and have driven the development of our ongoing modern civilization. Unlike natural disasters, environmental disasters are man-made, stemming from unawareness, ignorance, and illiteracy. This means that with appropriate measures, both their occurrence and impact can be reduced if not eliminated. Therefore, so far, the frequency and degree of environmental disaster damages have greatly decreased due to the development of science and technology and the strengthening of related laws and systems, mainly in advanced industrial countries.

A key point to note for the future is that humanity continuously creates new materials and products that can become sources of environmental disasters. In the past, humans used these innovations first, then realized the environmental problems caused by them and developed countermeasures. For example, it was a long time after combustible automobiles appeared in human life that photochemical smog was discovered and the need for countermeasures was raised. The same process of creating, using, finding solutions when problems occur, and then regulating new chemicals was repeated. Asbestos, DDT, and leaded gasoline are notable examples of this self-correcting process.

Humanity eventually created the most effective or optimal preventive measures after repeating such trials and errors. Current practices, mainly in advanced industrialized countries, include environmental impact assessments for any major developmental projects, ecological risk assessments for new chemicals, and life-cycle assessments for products. However, even with these regulations, risks are not entirely removed or zeroed-out, and environmental disasters causing numerous casualties continue to occur.

Until now, disaster prevention and response capabilities have greatly improved in advanced industrialized countries. More importantly, the common sense approach to share and transfer all such technologies and capabilities to other lesser and poorer societies and nations is truly achievable. So the future outlooks for environmental disasters may be optimistic. However, more sophisticated and careful systems are required because underdeveloped and developing countries are still very vulnerable to avoidable disasters, and advanced industrial countries can also experience disasters anytime, anywhere with small flaws and mistakes. The important tasks to execute and achieve in the future is to systematically study eco-disasters and educate the general public about related environmental knowledge. The numerous eco-disasters discussed in this book can serve as important case studies and a plentiful source of environmental knowledge in carrying out such tasks.

GPSR Compliance
The European Union's (EU) General Product Safety Regulation (GPSR) is a set of rules that requires consumer products to be safe and our obligations to ensure this.

If you have any concerns about our products, you can contact us on

ProductSafety@springernature.com

In case Publisher is established outside the EU, the EU authorized representative is:

Springer Nature Customer Service Center GmbH
Europaplatz 3
69115 Heidelberg, Germany

www.ingramcontent.com/pod-product-compliance
Lightning Source LLC
LaVergne TN
LVHW011802200926
848484LV00003B/763
9789819686667